MASTER EXAM WORKBOOK

Based on the National Electrical Code®

This workbook contains six closed book exams and ten open book exams for a total of over **1000** electrical exam questions with answers and references. This workbook is designed to help prepare the electrician for the Master electrical examination.

While every precaution has been taken in the preparation of this book, the author and publisher assumes no responsibility for errors or omissions. Neither is any liability assumed from the use of the information contained herein.

National Electrical Code® and NEC® are Registered Trademarks of the National Fire Protection Association, Inc., Quincy, MA.

ISBN 0 - 945495 - 18 - 8

Most electrical exams are given in three parts.

Part I Closed Book - 50 questions from practical knowledge. One hour time limit.

Part II Open Book - 70 questions. Two hour time limit.

Part III Open Book Calculations - 30 questions of calculations. Three hour time limit.

This Master exam workbook is designed to prepare the electrician for Part I and Part II of the exam.

PART I CLOSED BOOK EXAM
TIME LIMIT ONE HOUR   50 QUESTIONS

At 60 minutes you have 1.2 minutes (72 seconds) per question, which is plenty of time for a closed book type question. Simply read the question **carefully** and answer it. This part of the exam is a test of your knowledge in the electrical field. Questions are asked from theory, Ohm's law, safety, tools, and definitions from the NEC® Article 100.

Try to answer the 50 closed book questions from Part I in 30 minutes or less. At 30 minutes, you would have .6 minutes (36 seconds) per question which is plenty of time. Part I has a time limit of one hour, by working Part I in one-half hour you have an extra half-hour to apply to Part II of the exam. Part II open book consisting of 70 questions is the toughest of all three parts of the exam, so this is a great benefit to have an extra one-half hour.

PART II OPEN BOOK EXAM
TIME LIMIT TWO HOURS   70 QUESTIONS

At two hours (120 minutes) you have 1.7 minutes per question, which on some questions is **not** enough time. The key for more time is to work Part I closed book in 30 minutes or less, now you would have two and one-half hours or 150 minutes which now gives you 2.1 minutes per question for Part II open book.

Your score on Part II open book depends on how familiar you are with the Code. This is the toughest part of the exam trying to find the answers within the time format. Most students run out of time and never answer all of the questions in Part II. Of the 70 questions you'll be able to find 60 of them using the index in the Code book, you can figure on 10 questions that are difficult to find. The key to Part II open book is **not** to spend too much time on one question. If the question does not contain a key word in the index, **skip it**, and continue to the next question. Of the 70 questions you may skip 10 to 15 of them. Save these 10 or so questions for last, these 10 questions **won't fail you!** Find the other 60 questions and then answer the ones you skipped over. **Never** leave a question unanswered, unanswered is counted wrong. Always select a multiple choice answer.

After completing these practice exams turn to the answer sheets in the back of this book and grade yourself. 75% is passing.

To find your percentage simply divide the number of correct answers by the number of questions. Example: 38 correct answers divided by 50 questions would equal 76%.

I teach in my Electrical Code Classes that the key to the exam is that the student must first understand the question, which requires **careful reading of each word.**

As practice for the Open Book Exam Part II, try to find the correct answers for the following seven questions on the next page. Time limit for the 7 questions: 14 minutes. **GOOD LUCK!!**

After answering the seven open book questions turn to the answer section in the back of this book on page 158. This page contains the answers with references and the index key format.

I chose seven of the more difficult questions from the Masters exam so the student will realize how important **TIME** is in this part of the exam. Again, the key is don't spend too much time on a difficult to locate question, **skip** over it and move on, it won't fail you!

This workbook was designed to help you with this difficult area of the exam. Some students purchase a Code book just prior to taking an exam, and as you can see after completing this workbook, you are expected to be an expert in finding references in the Code book. How do you study a Code book for an exam? This workbook is the best way as it forces you to find the answers in the time limit. Write down your score and time spent on each exam in this workbook and notice your improvement as you work the latter exams.

To pass the exam its very simple, it takes work! Like with anything in life, you get out of it what you put into it. The more time you spend preparing for the exam the easier it will become.

As you work these exams and grade yourself, hi-lite the answers with a marking pen in your Code book.

OPEN BOOK PRACTICE   7 QUESTIONS   TIME LIMIT: 14 MINUTES

1. Circuits for lighting and power shall not be connected to any system containing trolley wires with a _____.

(a) busway   (b) ground return   (c) 600v system   (d) solid neutral

2. Branch circuits supplying only the ballasts for electric-discharge lamps mounted in permanently installed fixtures shall not exceed 600v between conductors and where installed in a tunnel the height shall not be less than _____ feet.

(a) 22   (b) 20   (c) 18   (d) 12

3. Which of the following letters indicate the quietest noise-rated ballast?

(a) A   (b) C   (c) D   (d) F

4. Proscenium light fixtures shall be wired with conductors having insulation suitable for not less than _____ degrees F.

(a) 86   (b) 30   (c) 104   (d) 257

5. Where auxiliary rectifiers are used with filter capacitors in the output for bias supplies, tube keyers, etc., bleeder resistors shall be used even though the DC voltage may not exceed _____ volts.

(a) 24   (b) 50   (c) 115   (d) 240

6. Storage batteries shall have leads that are covered with _____.

(a) nylon   (b) rubber   (c) thermoplastic   (d) (b) or (c)

7. Input leads to motor-generator or rotary converters shall be run separately from output leads.

(a) true   (b) false

CLOSED BOOK
EXAM
#1

50 QUESTIONS
TIME LIMIT - 1 HOUR

TIME SPENT [ ] **MINUTES**

SCORE [ ] **%**

MASTER CLOSED BOOK EXAM #1　　　| One Hour Time Limit |

1. If the phase voltages are 120 degrees apart, the line voltages will be _____ degrees apart.

(a) 30　(b) 60　(c) 90　(d) 120

2. Electrical current is measured in terms of _____.

(a) electrical pressure　　　　(b) electrons passing a point per second
(c) watts　　　　　　　　　　(d) ohms

3. A _____ is used for testing specific gravity.

(a) galvometer　(b) mandrel　(c) calorimeter　(d) hydrometer

4. In the course of normal operation the instrument which will be **least** effective in indicating that a generator may overheat because it is overloaded is _____.

(a) a stator thermocouple　　　　(b) a wattmeter
(c) a voltmeter　　　　　　　　　(d) an ammeter

5. An outlet box should be fastened to a solid concrete wall by means of _____.

(a) toggle bolts　(b) wood plugs　(c) lag bolts　(d) expansion bolts

6. What would be the advantage of 240 volt rather than 120 volt on a load with the same wattage?

(a) less power used　　　　(b) less voltage drop
(c) both (a) and (b)　　　　(d) neither (a) nor (b)

7. If frequency is constant, the inductive reactance of a circuit will _____.

(a) remain constant regardless of voltage or current change
(b) vary with voltage
(c) vary directly with current
(d) not effect the impedance

8. AC voltage of the system is **not** _____.

(a) EMF　(b) effective　(c) average　(d) RMS

9. A multimeter is a combination of _____.

(a) ammeter, ohmmeter and wattmeter
(b) voltmeter, ohmmeter and ammeter
(c) voltmeter, ammeter and megger
(d) voltmeter, wattmeter and ammeter

10. Two 120 volt light bulbs connected in series across 240 volt will _____.

(a) burn at full brightness   **(b) burn at half-brightness**
(c) burn out quickly   **(d) flicker with the cycle**

11. Ground-fault protection that functions to open the service disconnecting means _____ protect(s) service conductors or the service disconnecting means.

(a) adequately   (b) will   (c) will not   (d) totally

12. The conductor used to connect the grounded circuit of a wiring system to a grounding electrode is the _____.

(a) grounded conductor   **(b) bonding jumper**
(c) main bonding jumper   **(d) grounding conductor**

13. Galvanized rigid conduit is made of _____.

(a) iron   (b) zinc   (c) lead   (d) nickle

14. A soldering iron should not be heated to excess because it will usually _____.

(a) ruin the tin on the surface of the tip   **(b) burn the wooden handle**
(c) short-circuit the iron   **(d) anneal the copper tip**

15. A pendant fixture is a _____.

(a) hanging fixture   **(b) recessed fixture**
(c) bracket fixture   **(d) none of these**

16. _____ means so constructed or protected that exposure to the weather will not interfere with successful operation.

(a) Weatherproof   (b) Weather-tight   (c) Weather-resistant   (d) Weather-sealed

17. To reverse the rotation of a three-phase motor you would _____.

(a) turn it around
(b) reverse two of the four leads
(c) reverse any two of the three leads
(d) reverse all the leads

18. The disadvantage of AC compared to DC would be _____.

(a) smaller conductors to transmit electrical energy
(b) reduced cost of transmission with high voltage transformers
(c) less copper required with AC at higher voltages
(d) cannot be used in electro-plating

19. XL, XC, and R each in its own branch-circuit will flow in _____ directions.

(a) one   (b) two   (c) three   (d) none of these

20. In a highly inductive AC circuit, what device is used to correct the power factor towards unity?

(a) resistor   (b) inductor   (c) capacitor   (d) rectifier

21. In other than residential calculations, an ordinary outlet shall be calculated at _____.

(a) 660va   (b) 746w   (c) 2 amps   (d) 180va

22. New brushes for a generator should be fitted to the commutator by using a _____.

(a) bastard file
(b) strip of emery cloth
(c) polishing stone
(d) strip of sandpaper

23. Wound rotor and squirrel cage are two types of _____ motors.

(a) synchronous   (b) single-phase   (c) induction   (d) three-phase

24. A galvanometer is used to indicate _____.

(a) voltage   (b) current   (c) resistant   (d) wattage

25. The sum of the voltage drop around a circuit is equal to the source voltage is _A_.

(a) Kirchhoff's Law   (b) Ohm's Law   (c) Watt's law   (d) Nevin's Theory

26. Ohm's Law states _C_.

(a) E is equal to I/R   **(b) E is equal to R/I**
(c) E is equal to I times R   **(d) none of these**

27. A clamp-on ammeter will measure _d_.

(a) voltage when clamped on a single conductor
(b) current when clamped on a multiconductor cable
(c) accurately only when parallel to cable
(d) accurately only when clamped perpendicular to a conductor

28. The purpose of locknuts in making electrical connections on studs is to _A_.

(a) prevent the connection from loosening under vibration
(b) connect multiple conductors on the stud
(c) make the connection tamperproof
(d) avoid having to torque the studs

29. What winding of a current transformer will carry more current?

(a) primary   (b) secondary   (c) interwinding   (d) tertiary

30. The words "thermally protected" appearing on the nameplate of a motor indicates that the motor is provided with a _____.

(a) switch   (b) fuse   (c) breaker   (d) heat sensing element

31. What type of meter is shown in the diagram?

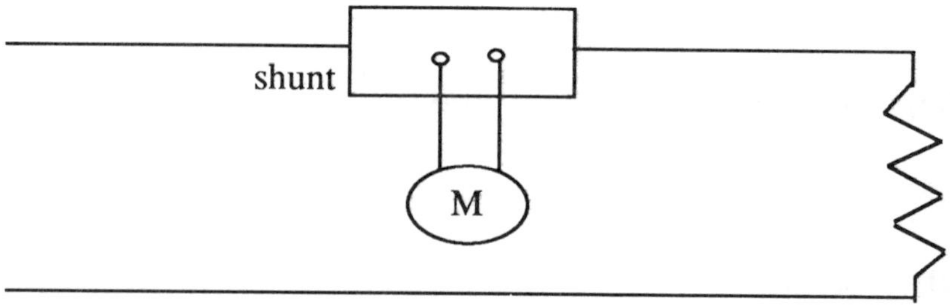

(a) wattmeter   (b) ammeter   (c) ohmmeter   (d) voltmeter

32. True power is always voltage times current _____ .

(a) in an AC circuit
(b) in a DC circuit
(c) where frequency is constant
(d) regardless of whether capacitive reactance is in the circuit or not

33. Which of the following would have the least amount of resistance?

(a) semi-conductor   (b) insulator   (c) resistor   (d) conductor

34. Grounding the metallic cover of flexible metal conduit and armored cable, is for protection against _____ .

(a) shock or injury   **(b) lightning**
(c) open field shorts   **(d) change of frequency**

35. Which of the following breaks down rubber insulation?

(a) oil   (b) acid   (c) water   (d) none of these

36. The opposite of resistance would be _____ .

(a) resonance   (b) reluctance   (c) reactance   (d) conductance

37. Conductors, such as non-metallic cable, when run in the space between studs are defined as _____ .

(a) inaccessible   (b) concealed   (c) enclosed   (d) buried

38. A type of transformer that has only a single winding is called a _____ transformer.

(a) tertiary   (b) auto   (c) isolation   (d) saturated

39. The main reason that water should not be used to extinguish fires involving electrical equipment is that water _____ .

(a) may transmit shock to the user
(b) is ineffective in extinguishing electrical fires
(c) may short out the equipment
(d) will ruin the electrical equipment

40. "Nichrome" wire is mainly used for _____.

(a) light bulb filaments   (b) heater coils
(c) motor windings   (d) an ammeter

41. In a multiple motor circuit there is feeder protection, branch-circuit protection and motor protection. If the feeder protection trips, the fault would be expected to be in _____.

(a) one of the motors   (b) one of the branch-circuits
(c) the feeder   (d) one of the starters

42. A branch-circuit that supplies a number of outlets for lighting and appliances is known as a _____ branch-circuit.

(a) general purpose   (b) multi-purpose   (c) utility   (d) none of these

43. The advantage of AC over DC includes which of the following?

(a) better speed control   (b) lower resistance at high currents
(c) ease of voltage variation   (d) impedance is greater

44. When three light bulbs are wired in a single fixture, they are connected in _____.

(a) parallel   (b) series   (c) series-parallel   (d) none of these

45. A voltage of 2000 microvolts is the same as _____ volts.

(a) 0.002   (b) 0.200   (c) 0.020   (d) 2,000

46. Electric motors are rated in terms of _____.

(a) dynes   (b) horsepower   (c) foot-pounds   (d) watts

47. _____ conductors shall be used for wiring on fixture chains.

(a) Solid   (b) Bare   (c) Covered   (d) Stranded

48. The proper way to mount an instrument meter is to _____.

(a) use a template   (b) use the meter to mark your holes
(c) drill from the back   (d) drill from the front

49. Three conductors (phase A, B and neutral) run from a busway to a load center should **not** be installed in three separate raceways because _____.

(a) you need room for future fill

(b) to avoid using a gutter

(c) of inductance

(d) its easier to pull the conductors

50. In a three-phase system with 4-wire, 208/120 volt, there exists _____.

(a) 3 ø 208 volt, 3 ø 120 volt, and 1 ø 120 volt
(b) 3 ø 208 volt, 1 ø 120 volt, and 1 ø 208 volt
(c) 3 ø 208 volt, 3 ø 120 volt, and 1 ø 208 volt
(d) 1 ø 208 volt, 3 ø 208 volt, and 3 ø 120 volt

CLOSED BOOK EXAM #2

50 QUESTIONS
TIME LIMIT - 1 HOUR

TIME SPENT ☐ **MINUTES**

SCORE ☐ %

MASTER CLOSED BOOK EXAM #2   One Hour Time Limit

1. Solenoids are made of what type of magnets?

(a) reverse   (b) permanent   (c) electro   (d) copper

2. An isolating switch is one that is _____.

(a) intended for cutting off an electrical circuit from its source of power
(b) required to have a padlock
(c) primarily used with an isolation transformer
(d) used only for heavy motor overloads

3. To measure the RPM speed of a motor you would use a _____.

(a) hydrometer   (b) bolometer   (c) tachometer   (d) odometer

4. Which of the following is **not** a factor in calculating the feeder conductor size?

(a) ambient temperature   (b) branch-circuit protection
(c) voltage drop   (d) demand factors

5. _____ is a synthetic non-flammable insulating liquid, which when decomposed by electrical arcs, involve non-flammable gases.

(a) Prynol   (b) Askarel   (c) Electrolyte   (d) Hermetic

6. Voltage drop in a conductor is _____.

(a) the conductor resistance times the voltage
(b) a function of insulation
(c) part of the load voltage
(d) a percentage of the applied voltage

7. The ratio of the maximum demand of the system to the total connected load of the system is called the _____ of the system.

(a) connected load   (b) nameplate   (c) demand factor   (d) turns-ratio

8. Piezoelectric is caused by crystals or binding _____.

(a) heat   (b) pressure   (c) static   (d) chemical

9. Since fuses are rated by amperes and voltage, a fuse will work on _____.

I. AC   II. DC   III. any voltage level

(a) I only   (b) II only   (c) III only   (d) either I or II

10. When three equal resistors are connected in parallel the total resistance is _____.

(a) the sum of the three resistors
(b) one-half of the sum of the three resistors
(c) less than any one resistor
(d) greater than any two equal resistors

11. Light fixtures hung by chains should be wired so that the _____.

(a) chain is not grounded   (b) wires support the light
(c) wires do not support the light   (d) light is insulated from the chain

12. _____ boxes may be weatherproof.

I. Rainproof   II. Raintight   III. Watertight

(a) I only   (b) II only   (c) III only   (d) I, II and III

13. The conductor insulation with the highest temperature rating would be _____.

(a) RUW   (b) THW   (c) RHH   (d) THWN

14. A switch or breaker should disconnect the grounded conductors of a circuit _____.

(a) only with an isolating switch
(b) before disconnecting the hot conductors
(c) simultaneously as it disconnects the ungrounded conductors
(d) the grounded conductor can never be switched

15. The definition of ambient temperature is _____.

(a) the temperature of the conductor
(b) the insulation rating of the conductor
(c) the temperature of the area surrounding the conductor
(d) the maximum heat the insulation can be used within

16. For voltage and current to be in phase ____.

(a) voltage and current appear at their zero and peak values at the same time
(b) the circuit impedance has only resistance
(c) neither (a) nor (b)
(d) both (a) and (b)

17. Which of the following instruments would you use to indicate the phase relationship between the voltage and the current of an AC circuit?

(a) KWH meter (b) power factor meter (c) manometer (d) growler

18. A ____ conductor is one having one or more layers of non-conducting materials that are not recognized as insulation.

(a) wrapped (b) bare (c) covered (d) insulated

19. Electrical equipment can be defined as ____.

I. fittings II. appliances III. devices IV. fixtures

(a) I only (b) I and IV only (c) I, III and IV (d) I, II, III and IV

20. The definition of a dry location ____.

(a) not normally subject to dampness
(b) not normally subjected to wetness
(c) may be temporarily subject to wetness
(d) all of the above

21. ____ duty is a type of service where both the load and the time intervals may have wide variations.

(a) Continuous (b) Periodic (c) Intermittent (d) Varying

22. An autotransformer is generally used rather than an isolation transformer ____.

(a) when cost is a factor
(b) where the ratio of transformation is low
(c) when you have several branch-circuits
(d) when safety is a factor

23. The symbol for a delta connection is _____.

(a) **Y** (b) **Ω** (c) **ø** (d) **Δ**

24. A _____ overcurrent device is not designated to interrupt short-circuit current.

(a) **inverse breaker**   (b) **dual-element fuse**   (c) **type S fuse**   (d) **thermal cutout**

25. If a test lamp lights when placed in series with a condenser and a suitable source of DC, it is a good indication that a condenser is _____ if the light burns continuously.

(a) **fully charged**   (b) **short circuited**
(c) **open circuit**   (d) **fully discharged**

26. Impedance is measured in _____.

(a) **farads** (b) **henrys** (c) **ohms** (d) **coloumbs**

27. The standard unit of electrical pressure is the _____.

(a) **ampere**   (b) **watt**   (c) **volt**   (d) **watt-hour**

28. The common fuse depends on the principle that the _____.

(a) **current flow developes heat**   (b) **overvoltage will expand the link**
(c) **increase of resistance will occur**   (d) **voltage developes heat**

29. The scope of the National Electrical Code includes _____.

(a) **conductors and equipment wherever and whenever electricity is employed**
(b) **all buildings or premises where electrical power is employed**
(c) **conductors and equipment on certain public or private premises**
(d) **all of these**

30. Electro-magnetic devices usually have a _____ core.

(a) **aluminum**   (b) **soft iron**   (c) **hard iron**   (d) **hard steel**

31. When a current leaves its intended path and returns to the source, bypassing the load, the circuit is _____.

(a) **broken**   (b) **open**   (c) **shorted**   (d) **incomplete**

32. The most heat is created when current flows through which of the following?

(a) a 10 ohm inductance coil
(b) a 10 ohm resistor
(c) a 10 ohm condenser
(d) the heat would be equal in all of these

33. The greater number of free electrons, the better the _____ of a metal.

(a) voltage drop   (b) resistance   (c) conductivity   (d) insulation value

34. When one material gives up electrons readily and another attracts them, this is the basis for _____.

(a) piezoelectric   (b) thermocouple   (c) electrochem   (d) photovoltaic

35. The voltage will lead the current when _____.

(a) capacitive reactance exceeds the inductive reactance in the circuit
(b) resistance exceeds reactance in the circuit
(c) reactance exceeds the resistance in the circuit
(d) inductive reactance exceeds the capacitive reactance in the circuit

36. The vector sum of the phase currents is equal to what in a balanced, resistive three-phase system?

(a) phase current x power factor   **(b) zero**
(c) 1.732 x phase current   **(d) three x phase current**

37. The heating of two different metals will cause _____.

(a) corrosion   (b) electron flow   (c) galvanic action   (d) fusion

38. If you are using two wattmeters to measure the power, you are measuring power in a _____ system.

(a) single-phase   (b) three-phase   (c) double   (d) high-voltage

39. When referring to a "8-32" machine bolt, the "32" refers to the _____.

(a) threads per inch   (b) length of the bolt   (c) diameter   (d) strength

40. On a DC ammeter, to increase the range you would use a ____.

(a) capacitor   (b) inductor   (c) shunt   (d) condenser

41. A motor works in the principles of ____.

(a) magnetism   (b) mechanical force   (c) residual force   (d) chemical action

42. Connecting batteries in series will give greater ____.

(a) amp hours   (b) amperage   (c) voltage   (d) all of these

43. What are incandescent light bulbs filled with?

(a) gas   (b) freon   (c) vacuum   (d) air

44. Root-mean-square is also called the ____ voltage.

(a) average   (b) peak   (c) effective   (d) maximum

45. The current through a DC arc lamp can be controlled by a ____.

(a) transformer   (b) rheostat   (c) rectifier   (d) scanner drum

46. The current used for charging storage batteries is ____.

(a) direct   (b) positive   (c) alternating   (d) negative

47. When threading rigid conduit the standard die should provide ____ taper per foot.

(a) 1/4"   (b) 1/2"   (c) 3/4"   (d) 1"

48. Lubrication would never be applied to a ____.

(a) bearing   (b) knife switch   (c) controller drum   (d) commutator

49. When the power factor in a given circuit is unity, the reactive power is ____.

(a) at maximum   (b) 1.1414   (c) zero   (d) a negative quantity

50. Materials containing numerous free electrons are ____.

(a) good insulators   (b) ferrous alloys   (c) good conductors   (d) carbons

CLOSED BOOK
EXAM
#3

50 QUESTIONS
TIME LIMIT - 1 HOUR

TIME SPENT [ ] **MINUTES**

SCORE [ ] **%**

1. The National Electrical Code is sponsored by the _____.

(a) **Underwriters Lab**
(b) **National Safety Council**
(c) **National Electrical Manufacturers Association**
(d) **National Fire Protection Association**

2. A drawing showing the floor arrangement of a building is referred to as a(an) _____.

(a) **perspective**   (b) **isometric**   (c) **surface G.B.**   (d) **plan**

3. Readily accessible is _____.

(a) **requiring only a 6 foot ladder**   (b) **requiring to move only small obstacles**
(c) **within 50 feet**   (d) **capable of being reached quickly for operation**

4. A Fitting is _____.

(a) **part of a wiring system that is intended primarily to perform an electrical function**
(b) **pulling cable into a confined area**
(c) **to be suitable or proper for**
(d) **part of a wiring system that is intended primarily to perform a mechanical function**

5. The neutral conductor _____.

(a) **is always the "white" grounded conductor**   (b) **carries the balanced current**
(c) **Note 8 never applies**   (d) **carries the unbalanced current**

6. An oil switch is a switch having contacts which operate under _____.

I. askarel   II. oil   III. other suitable liquids

(a) **I only**   (b) **II only**   (c) **I and II only**   (d) **I, II or III**

7. The definition of a qualified person is _____.

(a) **persons designated by the management of a building to supervise or work on equipment**
(b) **persons with a degree or license qualifying them to install, maintain, or operate a specific
     type of equipment or system**
(c) **a Master electrician**
(d) **persons familiar with the construction and operation of the equipment and the hazards
     involved**

8. All wiring must be installed so that when completed _____.

(a) it meets the current-carrying requirements of the load
(b) it is free of shorts and unintentional grounds
(c) it is acceptable to Code compliance authorities
(d) it will withstand a hy-pot test

9. If you were installing a fluorescent fixture in a customer's home and the noise emitted by the ballast were an important consideration, which one of the following sound ratings would you select to give the quietest operation?

(a) Type A   (b) Type D   (c) Type F   (d) Type AB

10. 60 cycle frequency travels 180 degrees in how many seconds?

(a) 1/60   (b) 1/120   (c) 1/180   (d) 1/30

11. A wattmeter indicates _____.

I. real power   II. apparent power if PF is not in unity   III. power factor

(a) I only   (b) II only   (c) III only   (d) I, II and III

12. The ratio of the maximum demand of a system, or part of a system, to the total connected load of a system or the part of the system under consideration is called the _____.

(a) load diversity   (b) demand ratio   (c) demand factor   (d) maximum demand

13. Code requirements may be waived _____.

(a) by the Fire department　　　　　**(b) by the utility company**
(c) by the authority having jurisdiction　　　**(d) under no circumstances**

14. Ampacity is _____.

(a) the current-carrying capacity of conductors expressed in volt-amps.
(b) the current-carrying capacity expressed in amperes.
(c) the current-carrying capacity of conductors expressed in amperes.
(d) the current in amperes a conductor can carry continuously under the conditions of use
　　　without exceeding its temperature rating.

15. An electric circuit that controls another circuit through a relay is a _____ circuit.

(a) remote-control   (b) pilot   (c) low-energy power   (d) transfer

16. Concrete, brick or tile walls are considered as being _____.

(a) isolated   (b) insulators   (c) grounded   (d) dry locations

17. Which of the following systems will furnish single-phase, three-phase, and two voltage levels?

(a) three-phase, 3-wire   **(b) three-phase, 4-wire**
(c) two-phase, 3-wire   **(d) single-phase, 4-wire**

18. All edges that are invisible should be represented in a drawing by lines that are _____.

(a) dotted   (b) curved   (c) solid   (d) broken

19. Covered conductor ampacity is _____ a TW insulated conductor of the same size.

(a) equal to   (b) greater than   (c) same as   (d) less than

20. An appliance that is not easily moved from one place to another in normal use is a _____ appliance.

(a) fastened in place   (b) dwelling-unit   (c) fixed   (d) stationary

21. A point on the wiring system at which current is taken to supply utilization equipment is called _____.

(a) a receptacle   (b) a junction   (c) an outlet   (d) a tap

22. _____ is a protective device for assembly as an integral part of a motor or motor-compressor and which, when properly applied, protects the motor against dangerous overheating due to overload and failure to start.

(a) thermal protector   **(b) thermal cutout**
(c) thermal overcurrent   **(d) thermal heat anticipator**

23. Listed equipment must be installed or used in accordance with _____.

(a) any instructions included in listing or labeling   **(b) the blue print**
(c) instructions provided by the manufacturer of the equipment   **(d) the electrician**

24. Compliance with the provisions of the Code will result in _____.

(a) freedom from hazard   **(b) an efficient system**
(c) good electrical service   **(d) convenient operation**

25. The horsepower rating of a motor _____.

(a) is a measure of motor efficiency   **(b) is the input to the motor**
(c) cannot be changed to watts   **(d) is the output of the motor**

26. What is the function of a neon glow tester?

I. determines if circuit is alive
II. determines polarity of DC circuits
III. determines if circuit is AC or DC current

(a) I only (b) II only (c) III only (d) I, II and III

27. Power in a three-phase system may be measured with a minimum of _____.

(a) one wattmeter (b) two voltmeters (c) two ammeters (d) none of these

28. What are Optical Fiber Cables used for?

I. transmit light for control II. signaling III. communications

(a) I only (b) II only (c) III only (d) I, II and III

29. What Chapter in the Code is power limited circuits referred to?

(a) Chapter 5 (b) Chapter 6 (c) Chapter 7 (d) Chapter 8

30. Something that would effect the ampacity of a conductor would be _____.

I. voltage II. amperage III. length IV. temperature

(a) I only (b) II only (c) III only (d) IV only

31. An electric bell outfit would be used to check for _____.

(a) voltage (b) ampacity (c) continuity (d) current

32. An ammeter is connected _____.

(a) in series with the load
(b) in parallel with the load
(c) both in series and in parallel with the load
(d) neither in series nor in parallel with the load

33. The Code requires all circuits to have at least _____ and adequate grounding.

(a) an ungrounded wire and a grounding wire   **(b) two hot wires and a grounding wire**
(c) at least two wires   **(d) a hot wire and a grounding wire**

34. Which of the following is **not** permitted for connecting conductors #6 and larger to terminal parts?

(a) solder lugs   (b) pressure connectors   (c) splices to flexible leads   (d) wire binding screws

35. _____ are used to provide explanatory material, references to other sections, precautions to be observed, and further mandatory requirements.

(a) mandatory rules   (b) exceptions   (c) fine print notes   (d) listings

36. Accessible (as applied to wiring methods) is _____.

I. not permanently closed in by the structure or finish of the building.
II. capable of being removed or exposed without damaging the building structure or finish.

(a) I only   (b) II only   (c) both I and II   (d) neither I nor II

37. Automatic is self-acting, operating by its own mechanism when actuated by some impersonal influence, such as a change in _____.

I. temperature   II. pressure   III. current strength

(a) I only   (b) I and II only   (c) II only   (d) I, II and III

38. Continuous duty is _____.

(a) a load where the maximum current is expected to continue for three hours or more.
(b) a load where the maximum current is expected to continue for one hour or more.
(c) intermittent operation in which the load conditions are regularly recurrent.
(d) operation at a substantially constant load for an indefinitely long time.

39. The definition of dustproof is _____.

(a) an electrical enclosure with sealing gaskets to prevent the equipment from overheating from an accumulation of dust.
(b) constructed so that dust will not enter the enclosing case under specified test conditions.
(c) protected so that dust will not interfere with its successful operation.
(d) a housing that seals dust from contacting energized parts.

40. The definition of identified is _____.

(a) materials included in a list published by an organization acceptable to the authority having jurisdiction and concerned with product evaluation.
(b) recognizable as suitable for the specific purpose, function, use, environment, etc., where described in a particular Code requirement.
(c) equipment to which has been attached a label, symbol, or other identifying mark of an organization acceptable to the authority having jurisdiction.
(d) acceptable to the authority having jurisdiction.

41. Covered, shielded, fenced or enclosed by means of suitable covers, casings, barriers, rails, screens, mats, or platforms is the definition of _____.

(a) guarded   (b) protected   (c) isolated   (d) enclosed

42. The overhead service conductors from the last pole or other aerial support to and including the splices, if any, connecting to the service entrance conductors at the building or other structure is the _____.

(a) service load   (b) service lateral   (c) service drop   (d) service supply

43. Without live parts exposed to a person on the operating side of the equipment is called _____.

(a) dead front   (b) isolated   (c) externally operable   (d) interrupted

44. Frequency is **not** a problem in most electrical work because it _____.

(a) never affects circuit values   (b) varies directly with voltage
(c) is a constant value   (d) only concerns DC

45. If the voltage is doubled, the ampacity of a conductor _____.

(a) increases   (b) decreases   (c) doubles   (d) remains the same

46. To calculate the power used by an appliance (not motor driven) one needs to know _____.

I. voltage and current   II. current and resistance   III. voltage and resistance

(a) I only   (b) II only   (c) III only   (d) I, II or III

47. Fixture wire shall not be smaller than # ____.

(a) 16   (b) 18   (c) 20   (d) 22

48. A toaster will produce less heat on low voltage because ____.

(a) its total watt output decreases   (b) the current increases
(c) the resistance has changed   (d) its total watt output increases

49. Electrical equipment shall be installed ____.

(a) better than the minimum Code allows
(b) according to the local Code when more stringent than the N.E.C.
(c) according to the N.E.C. regardless of the local Code
(d) according to the local Code when less stringent than the N.E.C.

50. The letters CATV refer to ____.

(a) communications (b) TV (c) antenna systems (d) video

CLOSED BOOK
EXAM
#4

50 QUESTIONS
TIME LIMIT - 1 HOUR

TIME SPENT [   ] **MINUTES**

SCORE [   ] **%**

1. It is good practice to use _____ in a long horizontal straight run of conduit.

(a) expansion joints   (b) flexible joints
(c) universal joints   (d) insulation joints

2. The ability of a device to open the maximum short or overload at the device, at a particular point in the electrical system is its _____ capacity.

(a) operating (b) interrupting (c) maximum (d) rated

3. To find the circular-mil area of a stranded conductor, the most accurate method is to _____.

(a) use a micrometer
(b) use a dial-o-meter
(c) square the number of strands
(d) multiply the number of strands by the circular-mil area of one strand

4. An AC ammeter is calibrated to read RMS values; this means the meter is reading the _____.

(a) maximum value (b) peak value (c) average value (d) effective value

5. Mica is commonly used for _____.

(a) pole insulators   (b) commutator bar separators
(c) panelboards   (d) appliance insulation

6. A hook on the end of a fish tape is **not** to _____.

(a) keep it from catching on joints and bends
(b) tie a swab to
(c) tie the wires to be pulled
(d) protect the end of the wire

7. Solders used for the connection of electrical conductors, are alloys of _____.

(a) tin and lead
(b) copper and tin
(c) zinc and lead
(d) zinc and tin

8. Which of the following statements is **not** an advantage of using an autotransformer?

(a) less expensive
(b) high efficiency
(c) better voltage regulation
(d) safe for stepping down higher voltages

9. _____ is/are the letter(s) used to indicate capactive reactance in a circuit.

(a) XL   (b) Z   (c) HZ   (d) XC

10. Utilization equipment uses _____.

(a) chemical or mechanical energy   **(b) electrical energy only**
(c) energy for electrical purposes   **(d) only mechanical energy**

11. A megger is used to measure _____.

(a) reactance   (b) specific gravity   (c) insulation resistance   (d) current

12. A _____ is a device which serves to govern in some predetermined manner the electric power delivered to the apparatus to which it is connected.

(a) switch   (b) feeder   (c) branch-circuit   (d) controller

13. Incandescent light bulbs when operated at a voltage less than the rated voltage, will normally result in _____.

(a) more light
(b) more current
(c) shorter bulb life
(d) longer bulb life

14. The volt-ampere rating in an AC circuit is a way to indicate the _____ power.

(a) true   (b) real   (c) apparent   (d) peak

15. In a three-phase, 4 wire wye system having 277v for lighting, the three-phase motors would be operating on a voltage of _____ volts.

(a) 208   (b) 480   (c) 240   (d) 190

16. Four heater coils with a given voltage will consume the most power when connected _____.

(a) all in series
(b) two in parallel
(c) all in parallel
(d) two parallel pair in series

17. A voltmeter measures _____.

(a) only voltage to ground (b) voltage difference
(c) AC voltage only (d) none of these

18. In a series circuit with four unknown resistive loads, it is certain that the _____.

(a) voltage drop across each load will be the same
(b) the current flowing through each load will be the same
(c) the wattage consumed by each load will be the same
(d) the total current in the circuit is the sum of all four loads

19. The lead covering on conductors is used _____.

(a) for grounding (b) for moisture proofing
(c) for easier soldering (d) to prevent vibration

20. The reason for rigid steel conduit to be galvanized or enameled is to _____.

(a) provide better grounding
(b) prevent corrosion
(c) make it easier to thread
(d) make painting easier

21. To check the power factor in an electrical load, if a power factor meter is not available, you would use _____.

(a) a wattmeter (b) an ammeter and voltmeter
(c) an ammeter, wattmeter and voltmeter (d) a KWH meter

22. A Plug fuse contains a safety element composed of a piece of metal that has a _____.

(a) low resistance and high melting point
(b) high melting point
(c) low resistance and low melting point
(d) low resistance

23. The definition of automatic is self-acting, operating by its own mechanism when actuated by some impersonal influence such as _____.

(a) a change in current strength   **(b) temperature**
(c) mechanical configuration   **(d) all of these**

24. DC voltage is reduced with a _____.

(a) resistor   (b) capacitor   (c) transformer   (d) condenser

25. The cord on a heating iron would most likely have an insulation of _____.

(a) nylon   (b) rubber   (c) asbestos   (d) cotton braid

26. If the clips on a cartridge fuse become hot, this is a good indication that _____.

(a) the fuse clips are too loose
(b) the voltage is too high
(c) the fuse clips are too tight
(d) the fuse rating is too high

27. A plastic bushing is considered a(n) _____.

(a) electrical device   (b) mounting device   (c) fitting   (d) insulated device

28. The current will lag the voltage when _____ is present in the circuit.

(a) capacitance   (b) inductance   (c) reluctance   (d) resistance

29. The voltage of a circuit is best defined as _____.

(a) the potential between two conductors
(b) the greatest difference of potential between two conductors
(c) the effective difference of potential between two conductors
(d) the average RMS difference of potential between any two conductors

30. The relationship of a transformer primary winding to the secondary winding is expressed in_____.

(a) wattage   (b) volt-amps   (c) turns-ratio   (d) amps

31. A hydrometer is used to measure _____.

(a) water currents   (b) water resistance   (c) specific gravity   (d) RPM

32. The least important reason for using a steel measuring tape around electrical equipment is
____.

(a) magnet effect
(b) danger of entanglement in rotating machinery
(c) shock hazard
(d) short circuit hazard

33. A megawatt is ____ watts.

(a) 100,000   (b) 1000   (c) 10,000   (d) 1,000,000

34. The resistance in a circuit may be increased by ____.

(a) increasing the current flow   **(b) increasing the EMF**
(c) a loose connection   **(d) decreasing the current flow**

35. D.P.D.T. letters would be used to identify a ____.

(a) circuit breaker   (b) type of insulation   (c) raceway   (d) switch

36. An accessible conductor is ____.

(a) not permanently enclosed by a structure
(b) admitting close approach
(c) not guarded by locked doors, elevation or other methods
(d) being reached without use of a ladder

37. The unit of measure for electrical capacity is the ____.

(a) ohm   (b) henry   (c) farad   (d) watt

38. For better illumination you would use ____.

(a) random spacing of lights   **(b) evenly spaced lights, higher ceilings**
(c) even spacing, numerous lights   **(d) cluster lights**

39. On a multi-wire, three-wire branch-circuit the maximum unbalanced load on the neutral
conductor at anytime would be when ____.

(a) the neutral is disconnected   **(b) both circuits are fully loaded**
(c) one hot leg is shut off   **(d) both circuits are off**

40. How many 4-way switches are required to switch a light from three places?

(a) one (b) two (c) three (d) four

41. _____ may be connected ahead of the main service switch.

(a) nothing (b) lighting fixtures (c) motors (d) lightning arresters

42. AC voltages may be increased or decreased by a _____.

(a) rectifier (b) motor (c) transformer (d) dynamo

43. The equipment used to measure and compare peak, average, and root-mean-square values on an AC voltage curve is called a _____.

(a) hydrometer (b) oscilloscope (c) thermal couple (d) mandrel

44. Which of the following is the best conductor of electricity?

(a) iron (b) aluminum (c) tungsten (d) carbon

45. Pure XL in a circuit will cause a _____ degree lag of the ampere curve to the voltage curve in an AC circuit.

(a) 90 (b) 120 (c) 180 (d) 240

46. _____ is/are theletter(s) used to indicate inductive reactance.

(a) XL (b) XC (c) HZ (d) Z

47. With respect to the safety value of insulation on electrical tools, it can be correctly stated that _____.

(a) the insulation should not be used as the only protective measure
(b) the insulation provides very little protection
(c) the insulation provides complete safety to the user
(d) insulation is really not needed

48. A DC voltmeter can be used directly to measure which of the following?

(a) cycles per second (b) power factor (c) power (d) polarity

49. A way of expressing the power factor is _____.

(a) θ (b) R/Z   (c) sin θ (d) XL/Z

50. Which of the following is best to fight electrical fires?

(a) soda-acid fire extinguisher   (b) fine spray of water
(c) foam fire extinguisher   (d) CO_2 fire extinguisher

CLOSED BOOK
EXAM
#5

50 QUESTIONS
TIME LIMIT - 1 HOUR

TIME SPENT [   ] MINUTES

SCORE [   ] %

1. The current will lead the voltage when _____.

(a) inductive reactance exceeds the capacitive reactance in the circuit
(b) reactance exceeds the resistance in the circuit
(c) resistance exceeds reactance in the circuit
(d) capacitive reactance exceeds the inductive reactance in the circuit

2. The difference between a thread on a bolt and a thread on a conduit is _____.

(a) the conduit thread is deeper
(b) the conduit thread is tapered
(c) no lubrication required in threading conduit
(d) the conduit thread always has the same pitch for any diameter

3. Open conductors run individually as service drops shall be _____.

I. insulated   II. bare   III. covered

(a) I only   (b) II only   (c) III only   (d) either I or III

4. A _____ is function that is similar to a rectifier.

(a) commutator   (b) transformer   (c) contactor   (d) inverter

5. The term pothead as used in the electrical trade means _____.

(a) current-limiting fuse   **(b) cable terminal**
(c) protective device   **(d) insulator on a line pole**

6. If an AC sine wave reaches a peak voltage of 100, what is the effective root-mean square voltage?

(a) 57.7 volts   (b) 141.4 volts   (c) 86.6 volts   (d) 70.7 volts

7. Raceways on the outside of buildings should be _____.

(a) weatherproof and covered
(b) watertight, arranged to drain
(c) raintight and arranged to drain
(d) rainproof and guarded

8. When measuring to determine the size of a stranded conductor, you would place the wire guage over ____.

(a) one strand of the conductor **(b) the insulation**
(c) all of the strands **(d) the outer covering**

9. The resistance of the filament in a light bulb is ____.

(a) usually the same at all times **(b) highest when bulb is off**
(c) lowest when bulb is on **(d) highest when bulb is on**

10. When one electrical circuit controls another electrical circuit through a relay the circuit is a ____ circuit.

(a) remote-control **(b) control** **(c) primary** **(d) secondary**

11. Two one ohm resistors in parallel, the total R is ____ ohm(s).

(a) 1 **(b) 2** **(c) 1/2** **(d) cannot be calculated**

12. The advantage(s) of a 4-wire, three-phase source over a 3-wire, three-phase source is/are ____.

I. two voltage levels   II. a grounded neutral   III. less copper required

(a) I only **(b) II only** **(c) III only** **(d) I and II**

13. An ammeter is disconnected from a current transformer (CT), the leads should be ____.

(a) taped **(b) shorted** **(c) re-routed** **(d) insulated**

14. When working on live 600 volt equipment where rubber gloves might be damaged, an electrician should ____.

(a) wear leather gloves over rubber ones
(b) work without gloves
(c) reinforce the fingers with tape
(d) carry a spare pair

15. A fuse puller is used to replace ____ fuses.

(a) cartridge **(b) plug** **(c) link** **(d) current-limiting**

16. Aluminum and copper-clad aluminum of the same circular-mil and insulation have ____.

(a) the same physical characteristics   **(b) the same termination methods**
(c) the same ampacity   **(d) different ampacities**

17. The primary purpose for grounding a raceway is to prevent the raceway from becoming ____.

(a) accidentally energized at a higher potential than ground
(b) a source of induction
(c) magnetized
(d) a path for eddy currents

18. When operated on a voltage 10% higher than nameplate rating, an appliance will:

I. have a shorter life   II. draw a higher current   III. use more power

(a) I only   (b) II only   (c) III only   (d) I, II and III

19. The liquid in a battery is called the ____.

(a) askarel   (b) eddy current   (c) hermetic   (d) electrolyte

20. ____ is current which flows through a circuit in the same direction at all times and with almost constant strength.

(a) AC   (b) Wattage   (c) DC   (d) Ampacity

21. An Erickson coupling is used to ____.

(a) join sections of EMT together
(b) connect EMT to flexible conduit
(c) connect two sections of rigid conduit when one section cannot be turned
(d) substitute for all-thread

22. To fasten a box to a terra cotta wall you would use ____.

(a) lag bolts   (b) expension bolts   (c) wooden plugs   (d) rawl plugs

23. If voltage rotation of a three-phase motor is A-B-C, the current rotation is ____.

(a) C-B-A   (b) A-C-B   (c) B-C-A   (d) B-A-C

24. The disadvantage of an autotransformer is _____.

(a) the high cost
(b) the poor efficiency due to a single winding
(c) the size
(d) the lack of isolation between the primary and secondary

25. The power factor can be determined by the use of which combination of the following meters?

I. ammeter   II. wattmeter   III. watt-hour meter   IV. ohmmeter
V. voltmeter   VI. phase rotation meter

(a) I, V and VI   (b) I, IV and VI   (c) I, II and V   (d) none of these

26. A unit of an electrical system which is intended to carry but not utilize electrical energy is a(an) _____.

(a) device   (b) motor   (c) light bulb   (d) appliance

27. When relay contacts make and break frequently, a condenser is used across the relay contacts to _____.

(a) energize the relay quicker   **(b) delay the coil energizing**
(c) reduce pitting of the contacts   **(d) increase the efficiency**

28. The main reason for using oil in an oil circuit breaker is to _____.

(a) lubricate the points   **(b) quench the arc**
(c) increase the capacity of current   **(d) decrease the resistance**

29. The resistance of a copper conductor is _____.

(a) inversely proportional to its cross-sectional area
(b) inversely proportional to its length
(c) directly proportional to its diameter
(d) directly proportional to the square of its diameter

30. To measure AC cycles per second, you would use a _____.

(a) hydrometer   (b) manometer   (c) frequency meter   (d) power factor meter

31. Voltaic current will lead the voltage when _____.

(a) chemical   (b) solar   (c) both of these   (d) none of these

32. In AC circuits, inductance is measured in _____.

(a) farads   (b) henrys   (c) coulombs   (d) ohms

33. A motor with a wide speed range is a(n) _____.

(a) DC motor   (b) AC motor   (c) synchronous motor   (d) induction motor

34. For discharge lighting lamps to operate properly they would require _____.

(a) some sort of ballast   (b) a starter
(c) resistive starter   (d) a cool ambient

35. Comparing incandescent lighting with fluorescent lighting at the same level of illumination, the cost of energy for fluorescent would be _____.

(a) greater   (b) less   (c) the same   (d) 1.05 increase

36. Motor horsepower ratings are based on an observable safe temperature rise above ambient temperature. The ambient temperature is taken as _____ degrees C.

(a) 40   (b) 45   (c) 50   (d) 60

37. In a three-phase circuit, the legs of the phase are _____ electrical degrees apart.

(a) 90   (b) 180   (c) 120   (d) 240

38. A generator works on the basis of _____.

(a) a conductor moving inside a magnetic field   (b) ozone principle
(c) a magnetic field moving around a conductor   (d) none of these

39. In an area that requires explosion-proof wiring, raceways entering a box in this area which contains equipment that may produce sparks, shall be provided with _____.

(a) an approved sealing compound   (b) insulated bushings
(c) two grounding conductors   (d) triple lock nuts

40. In a 60 cycle system, what length of time does it take to go 90 degrees?

(a) 1/3 second   (b) 1/90 second   (c) 1/60 second   (d) 1/240 second

41. The basic unit of electrical work is the _____.

(a) volt-amp   (b) watt   (c) watt-hour   (d) kva

42. A good insulator for very high temperature is _____.

(a) mica   (b) rubber   (c) plastic   (d) bakelite

43. A wattmeter is connected in _____ in the circuit.

(a) series   (b) parallel   (c) series-parallel   (d) none of these

44. The main reason of laminating the core of a power transformer is to keep the _____.

(a) copper losses at a minimum   (b) turns-ratio balanced
(c) eddy current loss at a minimum   (d) friction losses low

45. To open a single-throw switch, the switch should be mounted so that the switch blade must move _____.

(a) downward   (b) upward   (c) twice   (d) slowly

46. The total opposition to current flow in an AC circuit is expressed in ohms as _____.

(a) impedance   (b) conductance   (c) reluctance   (d) resistance

47. Which of the following motors does **not** have a commutator?

(a) squirrel cage   (b) series   (c) shunt   (d) synchronous

48. Resistance can be found directly from wattage and voltage measurements by the equation _____.

(a) $R = \dfrac{V^2}{W}$ (b) $R = \dfrac{V}{W}$ (c) $R = \dfrac{W}{V}$ (d) $R = \dfrac{W^2}{V}$

49. A coil has 100 turns. A current of 10 amps is passed through the coil. The coil developes
_____.

(a) 1000 watts   (b) 1 kva   (c) mutual inductance   (d) 1000 amp turns

50. A substance that would be good as an electrical insulation is which of the following?

(a) carbon   (b) oil   (c) lead   (d) iron

CLOSED BOOK EXAM #6

50 QUESTIONS
TIME LIMIT - 1 HOUR

TIME SPENT [ ] MINUTES

SCORE [ ] %

1. A high spot temperature in a corroded electrical connection is caused by a(an) _____.

(a) increase in the flow of current through the connection
(b) decrease in the voltage drop across the connection
(c) increase in the voltage drop across the connection
(d) decrease in the effective resistance of the connection

2. A hickey is ____.

(a) a tool used to bend small sizes of rigid conduit   (b) a part of a conduit
(c) not used in the electrical trade   (d) used only by a plumber

3. The lubricant used for a motor sleeve bearing would be ____.

(a) vaseline (b) grease (c) oil (d) graphite

4. The resistance of a circuit may vary due to ____.

(a) a loose connection (b) change in voltage (c) change in current (d) induction

5.The service disconnecting means for each set of service-entrance conductors shall consist of not more than ____ switches.

(a) 6 (b) 8 (c) 10 (d) 12

6. A light bulb usually contains ____.

(a) air (b) neon (c) H2O (d) either a vacuum or gas

7. A ____ is an enclosure designed either for surface or flush mounting and provided with a frame, mat, or trim in which a swinging door or doors are or may be hung.

(a) cabinet (b) panelboard (c) cutout box (d) switchboard

8. ____ is a system or circuit conductor that is intentionally grounded.

(a) grounding conductor   (b) grounded conductor
(c) neutral conductor   (d) grounding electrode conductor

9. The definition of overload is ____.

(a) the operation of equipment in excess of normal full-load rating, or of a conductor in excess of rated ampacity.
(b) a thermal relay used to protect a motor.
(c) any current in excess of the rated current of equipment or the ampacity of a conductor.
(d) the heaters in a motor circuit.

10. A requirement of service that demands operation for alternate intervals of (1) load and no load; or (2) load and rest; or (3) load, no load, and rest is called ____ duty.

(a) variable (b) intermittent (c) short-time (d) periodic

11. An outlet where one or more receptacles are installed is called a ____.

(a) multi-outlet assembly (b) receptacle outlet (c) duplex outlet (d) tri-plex outlet

12. A conductor encased within material of composition or thickness not recognized by the Code is a ____ conductor.

(a) coated (b) semi (c) covered (d) fiber optic

13. Cooling of electrical equipment within enclosures is ____.

(a) the responsibility of the equipment manufacturer   (b) not covered by the Code
(c) covered by the Code   (d) not required

14. Approved is ____.

(a) listed and labeled equipment
(b) acceptable to the authority having jurisdiction
(c) tested and approved for the purpose by a qualified testing lab
(d) UL listed only

15. A branch circuit is ____.

(a) the circuit conductors between the final overcurrent device protecting the circuit.
(b) the circuit conductors between the final overload device protecting the circuit.
(c) the circuit conductors prior to the final overcurrent device protecting the circuit.
(d) the conductors between the service and the sub-panel.

16. What chapter in the Code is Mobile Homes referred to?

(a) Chapter 3 (b) Chapter 5 (c) Chapter 8 (d) Chapter 9

17. Is it permissible to install direct current and alternating current conductors in one pull box?

(a) Yes, if insulated for the maximum voltage of any conductor.
(b) No, never.
(c) Yes, if ampacity is the same for both conductors.
(d) Yes, in dry places.

18. The maximum size fuse to be used in a branch circuit containing no motors depends on the _____.

(a) load   (b) wire size   (c) voltage   (d) switch size

19. The type letter for moisture and heat resistant rubber is _____.

(a) RUH   (b) THW   (c) RHW   (d) MHR

20. Copper-clad aluminum conductors form a minimum of _____ percent of copper of the cross-section area of a solid conductor or each strand of a stranded conductor.

(a) 10   (b) 15   (c) 20   (d) 25

21. A current-limiting overcurrent protective device is a device which will _____ the current flowing in the faulted circuit.

(a) reduce   (b) increase   (c) maintain   (d) none of these

22. The conductors and equipment for delivering energy from the electricity supply system to the wiring system of the premises served is called the ______.

(a) primary   (b) distribution   (c) main supply feeder   (d) service

23. Concealed is _____.

(a) not readily visible   (b) made inaccessible by the structure or finish of the building
(c) surrounded by walls   (d) attached to the surface

24. Continuous load is _____.

(a) a load where the maximum current is expected to continue for three hours or more.
(b) a load where the maximum current is expected to continue for one hour or more.
(c) intermittent operation in which the load conditions are regularly recurrent.
(d) operation at a substantially constant load for an idefinitely long time.

25. If a live conductor is contacted accidentally, the severity of the electrical shock is determined primarily by _____.

(a) whether the current is AC or DC   **(b) the current in the conductor**
(c) the size of the conductor   **(d) the contact resistance**

26. Electron flow produced by means of applying a pressure to a material is called _____.

(a) thermoelectricity   (b) piezoelectricity   (c) electrochemistry   (d) photo conduction

27. A dual voltage motor will run more efficiently _____.

(a) at the lower voltage   **(b) at the higher voltage**
(c) the same at either voltage   **(d) none of these**

28. The term anode refers to _____.

(a) capacitor   (b) dynamo   (c) rectifier   (d) inductor

29. A box contains one grounded and three ungrounded conductors, from one ungrounded to the grounded conductor 208 volts is measured, the other two ungrounded measure 120 volts to the grounded conductor, the system is _____.

(a) delta   (b) wye 3-wire   (c) wye 4-wire   (d) open delta 4-wire

30. The voltage drop in a line can be decreased by _____.

I.   increasing the resistance   II. increasing the current
III. decreasing the load   IV. increasing the wire size

(a) III only   (b) I and III only   (c) III and IV only   (d) I, III and IV only

31. A 10 ohm resistance carrying 10 amperes of current uses _____ watts of power.

(a) 100   (b) 20   (c) 500   (d) 1000

32. When using a #12-2 with ground, the ground _____ carry current under normal operation.

(a) will   (b) will not   (c) will sometimes   (d) none of these

33. A transformer is more efficiently utilized when the load has a ____ power factor.

(a) low   (b) medium   (c) average   (d) high

34. When re-routing conduit, it may be necessary to increase the wire size if the distance is considerably greater, in order ____.

(a) to account for current drop   **(b) to allow for possible resistance drop**
(c) to compensate for voltage drop   **(d) to increase the mechanical strength**

35. To handle a three-phase unbalanced system, balance the system by making all loads equal to the ____ single phase load.

(a) smallest   (b) average   (c) largest   (d) unbalanced

36. The connection between the grounded circuit conductor and the equipment grounding conductor at the service is called the ____.

(a) equipment bonding jumper   **(b) main bonding jumper**
(c) circuit bonding jumper   **(d) electrode bonding jumper**

37. Connections of conductors to terminal parts must ensure ____.

(a) a good connection without damaging the conductors
(b) proper torque values
(c) proper bonding
(d) proper crimping pressure

38. To determine directly whether all finished wire installations possess resistance between conductors, and between conductors and ground, use ____.

(a) set screws   (b) shields   (c) clamps   (d) a megger

39. Most electric power tools, such as electric drills, come with a third conductor in the power lead which is used to connect the case of the tool to a grounded part of the electric outlet. The purpose for having this extra conductor is to ____.

(a) protect the user of the tool should the winding break down to the case.
(b) eliminate sparking between the tool and the material being worked upon.
(c) provide for continued operation of the tool should the regular grounded line-wire open.
(d) prevent accumulation of a static charge on the case.

40. The purpose of having a rheostat in the field of a DC shunt motor is to _____.

(a) control the speed of the motor
(b) minimize the starting current
(c) limit the field current to a safe value
(d) reduce sparking at the brushes

41. You are to check the power factor of a load, you cannot get a power factor meter, you would use _____.

(a) a wattmeter
(b) a voltmeter and an ammeter
(c) a kilo-watt hour meter
(d) an ammeter, a wattmeter and a voltmeter

42. The factor which will have the **least** effect on the voltage at the most distant point from the source of supply for a two wire circuit is _____.

(a) whether the supply is 25 cycle or 60 cycle AC
(b) the gauge of the circuit wires
(c) the amount of load on the circuit
(d) the length of the circuit

43. The load side is usually wired to the blades of a knife switch to _____.

(a) prevent blowing the fuse when opening the switch.
(b) make the blades dead when the switch is opened.
(c) prevent arcing when the switch is opened.
(d) allow changing of fuses without opening the switch.

44. High-voltage cable which is to be installed in underground ducts is generally protected with a _____.

(a) copper outer jacket   (b) lead sheath   (c) steel wire armor   (d) tarred jute covering

45. A conduit coupling is sometimes tightened by using a strap wrench rather than a Stillson wrench, the strap wrench is used when it is important to avoid _____.

(a) crushing the conduit
(b) bending the conduit
(c) stripping the threads
(d) damaging the outside finish

46. The life of insulation used in electrical installations is directly affected by heat. Of the following, the electrical insulation which can **least** withstand heat is _____.

(a) mica   (b) rubber   (c) fiberglass   (d) enamel

47. With respect to a common light bulb, it is correct to state that the _____.

(a) circuit voltage has no effect on the life of the bulb.
(b) base has a left hand thread.
(c) filament is made of carbon.
(d) lower wattage bulb has the higher resistance.

48. One identifying feature of a squirrel-cage induction motor is that it has no _____.

(a) air gap
(b) commutator or slip rings
(c) iron core in the rotating part
(d) windings on the stationary part

49. If the allowable current in a copper bus bar is 1000 amperes p.s.i. of cross-section, the width of a standard 1/4" bus bar designed to carry 1500 amperes would be _____.

(a) 2"   (b) 4"   (c) 6"   (d) 8"

50. The NEC requires that conduit must be continuous from outlet to outlet, must be mechanically and electrically connected to all fittings, and must be suitably grounded. The reason for having the conduit electrically continuous and grounded is to _____.

(a) shield the wires inside the conduit from external magnetic fields.
(b) provide a metallic return conductor.
(c) make it easy to test wiring connections.
(d) prevent electrical shock which might otherwise result from contact with the conduit.

OPEN BOOK
EXAM
#1

70 QUESTIONS
TIME LIMIT - 2 HOURS

TIME SPENT [   ] MINUTES

SCORE [   ] %

1. In commercial garages, generally the floor area to a height of 18" above grade is designated as _____.

(a) Class I, Division I **(b) Class I, Division II**
(c) Class II, Division II **(d) Class II, Division I**

2. In using multiple grounding electrodes, they shall be separated one from the other at _____ feet distance apart.

(a) 6 (b) 8 (c) 10 (d) 12

3. The disconnecting means for a 50 hp three-phase 460v induction motor shall have an ampere rating of at least _____ amps.

(a) 126 (b) 75 (c) 91 (d) 63

4. In combustible walls or ceilings, the front edge of an outlet box or fitting may set back of the finished surface _____.

(a) 1/4" (b) 1/8" (c) 1/2" (d) not at all

5. Overcurrent protection for electric organ circuits shall not exceed _____ amps.

(a) 15 (b) 20 (c) 25 (d) none of these

6. Each transformer shall be provided with a nameplate giving the name of the manufacturer, rated kva, _____ if 25 kva and larger.

I.   insulation type
II. frequency
III. impedance
IV. amount and kind of insulating liquid if used

(a) I only (b) II only (c) IV only (d) I, II, III and IV

7. A single receptacle installed on an individual branch circuit shall have a rating not less than _____ percent of the rating of the branch circuit.

(a) 50 (b) 80 (c) 100 (d) 125

8. In pull boxes or junction boxes having any dimension over _____ feet, all conductors shall be cabled or racked up in an approved manner.

(a) 2   (b) 3   (c) 4   (d) 6

9. A surge arrester is a protective device for limiting surge voltages by _____ or bypassing surge current.

(a) decreasing   (b) discharging   (c) limiting   (d) derating

10. Conductors to the hoistway door interlocks from the hoistway riser shall be _____.

(a) flame retardant   (b) rated 75 degrees C minimum
(c) rated 200 degrees C   (d) (a) and (c)

11. Flexible metal conduit used as a fixture whip to bond a fixture enclosure to a circuit junction box must not exceed _____ feet in length.

(a) 4   (b) 8   (c) 6   (d) 10

12. A disconnecting means serving a hermetic refrigerant motor compressor selected on the basis of the nameplate rated load current or branch circuit selection current, whichever is greater shall have an ampere rating of _____ percent of the nameplate rated load current or branch circuit selection current.

(a) 125   (b) 80   (c) 100   (d) 115

13. What size grounding conductor is required for a two-wire DC generator used in conjunction with balancer sets to obtain neutrals for 3-wire system equipped with overcurrent devices that will disconnect the 3-wire system in case of excessive unbalancing of voltages or currents?

(a) #10 solid copper minimum size
(b) shall not be smaller than any system conductor
(c) shall not be smaller than the neutral conductor
(d) sized from Table 250-94

14. The hazardous area in a pit of a spray operation without proper ventilation is classified as a _____ location.

(a) Class I,   Division I
(b) Class I,   Division II
(c) Class II,   Division I
(d) Class III, Division I

15. Temporary electrical power and lighting installations shall be permitted for a period not to exceed _____ days.

(a) 90 (b) 60 (c) 30 (d) 15

16. When an outlet from an underfloor raceway is discontinued, the circuit conductors supplying the outlet _____.

(a) may be spliced
(b) may be reinsulated
(c) may be handled like abandoned outlets on loop wiring
(d) shall be removed from the raceway

17. Emergency power panel conductors supplying a building are tapped on _____.

(a) the line side of the service
(b) any sub-feed panel
(c) any feeder circuit
(d) any circuit breaker main

18. Fixtures shall be wired with conductors having insulation suitable for the environment conditions and _____ to which the conductors will be subjected.

(a) temperature (b) voltage (c) current (d) all of these

19. Isolated non-carrying metal parts of outline lighting may be bonded by a _____ conductor.

(a) #14 (b) #12 (c) #10 (d) #8

20. Solid dielectric insulated conductors operated above 2000 volts in permanent installations shall have _____ insulation and shall be shielded.

(a) ozone-resistant b asbestos (c) hi-temperature (d) perfluoro-alkoxy

21. The grounding conductor shall be identified by _____.

(a) one continuous green color   **(b) being bare**
(c) a continuous green color with yellow stripes   **(d) any of these**

22. In general, the voltage limitation between conductors in surface metal raceway is _____ volts.

(a) 300 (b) 600 (c) 900 (d) 1000

23. A disconnecting means for X-ray equipment shall have adequate capacity for at least _____.

(a) 50% of the input required for momentary rating of equipment
(b) 100% of the input required for momentary rating of equipment
(c) 50% of the input requirement for long time rating
(d) 125% of the input requirement for long time rating

24. Locations where conbustible fibers are stored are designated as _____.

(a) Class II, Division II
(b) Class III, Division I
(c) Class III, Division II
(d) non-hazardous

25. A switch or circuit breaker should disconnect the grounded conductors of a circuit _____.

(a) by hand levers only
(b) simultaneously as it disconnects the ungrounded conductors
(c) before it disconnects the ungrounded conductors
(d) in none of the above ways

26. The space measured horizontally above a show window must have at least one receptacle for each _____ linear feet.

(a) 8 (b) 10 (c) 12 (d) 6

27. Wires run above heated ceilings and within thermal insulation, shall be derated on the basis of _____ degrees C.

(a) 86 (b) 30 (c) 50 (d) 20

28. If the allowable current carrying capacity of a conductor does not correspond to the rating of a standard size overcurrent device, the next larger size may be used provided the current does not exceed _____ amps.

(a) 200 (b) 500 (c) 800 (d) 1000

29. It is recognized that _____ may be desirable for service disconnecting means rated less than 1000 amperes on solidly grounded systems having more than 150 volts to gound not exceeding 600 volts phase to phase.

(a) current limiting   **(b) fuses**
(c) ground fault protection   **(d) none of these**

30. Flourescent lighting fixtures may be used as raceways if _____.

(a) they are connected by a conduit wiring method
(b) they are wired so that conductors are not closer than 3" from the ballast
(c) approved for use as a raceway
(d) none of these

31. The feeder for six 20 amp receptacles supplying shore power shall be calculated at _____ percent of the sum of the rating of the receptacles.

(a) 70   (b) 80   (c) 90   (d) 100

32. _____ conductors shall be used for wiring on fixture chains and other movable parts.

(a) Solid   (b) Covered   (c) Insulated   (d) Stranded

33. The rating of a lampholder on a circuit which operates at a voltage less than 50 volts shall be at least _____ watts.

(a) 220   (b) 660   (c) 330   (d) 550

34. Health care low voltage equipment frequently in contact with bodies of persons shall not exceed _____ volts.

(a) 50   (b) 115   (c) 10   (d) 8

35. Where oil switches, air or oil circuit breakers constitute the service disconnecting means for over 600 volts, a(n) _____ isolating switch shall be installed on the supply side of the disconnecting means.

(a) air break   (b) knife blade   (c) position   (d) double-throw

36. Rigid metal conduit shall be permitted to be installed in concrete, in direct contact with the earth, or in areas subject to severe influences where protected by _____ and judged suitable for the condition.

(a) ceramic   (b) corrosion protection   (c) PVC   (d) orangeburg

37. In a mobile home, at least how many inches of free conductor shall be left at each outlet?

(a) 3"   (b) 4"   (c) 5"   (d) 6"

38. Nonmetallic extensions shall be supported every _____ inches.

(a) 6   (b) 8   (c) 10   (d) 16

39. The National Electrical Code _____.

(a) is intended to be a design manual
(b) is meant to be used as an instruction annual for untrained persons
(c) is intended to protect persons and property
(d) is published by Bureau of Standards

40. Which of the following is/are correct about open wire systems on insulators?

I. surface-type snap switches do not need boxes
II. conductor supports shall be within 6" of a tap
III. surface-type snap switches shall be mounted on insulating material

(a) I only   (b) II only   (c) I and III only   (d) I, II and III

41. Which of the follwoing is not a type of optical fiber cable?

(a) AC   (b) nonconductive   (c) conductive   (d) hybrid

42. The equipment bonding jumper on the supply side of the service is sized by the rating of the _____.

(a) overcurrent protective device
(b) service entrance conductors
(c) service drop
(d) load to be served

43. Transformers over 112 1/2 kva shall not be within _____ of combustible material.

(a) 12"   (b) 16"   (c) 24"   (d) 30"

44. How many #12 XHHW conductors can you install in a 3/8" flexible metal conduit with the use of inside fittings?

(a) 1   (b) 2   (c) 3   (d) 0

45. The overall covering of UF cable shall be _____.

I. suitable for burial
II. flame- retardant
III. moisture, fungus and corrosion resistant

(a) III and II only   (b) I only   (c) I and III only   (d) I, II and III

46. In a Class II location, where electrically-conducting dust is present, flexible connections at motors could be made with _____.

(a) flexible metal conduit
(b) type AC armored cable
(c) hard usage cable
(d) liquid-tight flexible metal conduit with approved fittings

47. The short-circuit and ground-fault protective device protecting the branch circuit shall have sufficient _____ to permit the motor-compressor to start.

(a) voltage   (b) current   (c) time delay   (d) capacity

48. Where magnesium, aluminum or aluminum-bronze powders may be present, transformers _____.

(a) must be dust-tight
(b) may be pipe-ventilated
(c) must be approved for Class II, Division I
(d) are not allowed

49. Knife switches rated for more than 1200 amperes at 250 volts _____.

(a) are used only as isolating switches
(b) may be opened under load
(c) should be placed so that gravity tends to close them
(d) should be connected in parallel

50. All electric equipment, including power supply cords, used with storable pools shall be protected by _____.

(a) fuses   (b) circuit breakers   (c) double-insulation   (d) GFCI

51. What is the minimum working clearance on a circuit 120 volts to ground, exposed live parts on one side and no live or grounded parts on the other side of the working space?

(a) 3 feet   (b) 3 1/2 feet   (c) 4 feet   (d) 6 feet

52. Where the calculated number of conductors, all of the same size, includes a decimal fraction, the next higher whole number shall be used where this decimal is _____ or larger.

(a) .5   (b) .08   (c) .008   (d) .8

53. An oil switch is used on a motor circuit whose rating _____ or 100 amps.

(a) is 600 volt   **(b) exceeds 600 volts**
(c) does not exceed 600 volts   **(d) none of these**

54. The conductor between a lightning arrester and the line for installations operating at 1000 volts or more must be at least _____ copper.

(a) #14   (b) #6   (c) #8   (d) #12

55. Electric sign enclosures may be constructed of wood if _____.

(a) kept 1" from lampholders   **(b) wood is not permitted at all**
(c) kept 2" from lampholders   **(d) none of these**

56. If the maximum value of an AC current is 50 amps, the R.M.S. value would be approximately _____ amps.

(a) 25   (b) 30   (c) 35   (d) 50

57. Fuses, circuit breakers, or combinations thereof, shall not be connected in parallel except _____.

I. field installed   II. factory assembled   III. readily accessible

(a) I only   (b) II only   (c) III only   (d) I and III

58. Electric-discharge tubing operating at more than 7500 volts shall clear the adjacent surface by not less than _____.

(a) 1/4"   (b) 1/2"   (c) 1"   (d) 3"

59. The ampacity of the phase conductors from the generator terminals to the first overcurrent device shall not be less than _____ percent of the nameplate rating of the generator.

(a) 75   (b) 115   (c) 125   (d) 140

60. Flexible cord shall not be used as a substitute for _____ wiring.

(a) temporary   (b) fixed   (c) concealed   (d) none of these

61. The smallest size diameter pipe that can be used for a made electrode is _____.

(a) 1/2"   (b) 5/8"   (c) 3/4"   (d) 1"

62. A night club lighting dimmer installed in an ungrounded conductor shall have overcurrent protection rated at no more than ____ percent.

(a) 50 (b) 70 (c) 80 (d) 125

63. A motel conference room is designed for the assembly of 100 or more persons. The room is fire-rated construction. One of the following wiring methods shall be required:

(a) rigid nonmetallic conduit
(b) MI cable
(c) nonmetallic sheathed cable
(d) type AC cable

64.Busways rated over 600 volts shall have all connection hardware accessible for ____.

(a) installation (b) connection (c) maintenance (d) all of these

65. In areas where walls are frequently washed, conduit should be mounted with ____ inches of air space between the wall and the conduit.

(a) 1/8" (b) 1/4" (c) 3/8" (d) 1/2"

66. #12 aluminum conductors completely enclosed in the motor control enclosure shall be permitted to be protected with an overcurrent device where the rating is not more than ____ amperes.

(a) 20 (b) 80 (c) 100 (d) 125

67.Cable tray systems shall not be used in ____ or where subject to damage.

(a) tunnels (b) hoistways (c) hazardous locations (d) 600 volt systems

68. What is the minimum burial depth for rigid nonmetallic conduit in a dispensing station Class I, Division I location?

(a) 18" (b) 24" (c) 30" (d) cannot be used in Class I, Division I location

69. Conductors ____ and larger shall be stranded when installed in raceways.

(a) #12 (b) #10 (c) #8 (d) none of these

70. The minimum size service for a mobile home in a mobile home park is ____ amp.

(a) 80 (b) 70 (c) 200 (d) 100

OPEN BOOK
EXAM
#2

70 QUESTIONS
TIME LIMIT - 2 HOURS

TIME SPENT [ ] **MINUTES**

SCORE [ ] **%**

1. Tap conductors in a metal raceway for recessed fixture connections shall be limited to _____ feet in length.

(a) 2   (b) 4   (c) 6   (d) 10

2. The number of square feet that each made plate electrode should present to the soil is _____.

(a) 4 square feet   (b) 3 square feet   (c) 2 square feet   (d) 1 square foot

3. An autotransformer starter shall provide _____.

I.   an "off position"
II. a running position
III. at least one starting position

(a) I only   (b) II only   (c) I and II only   (d) I, II and III

4. When service-entrance conductors exceed 1100 MCM for copper, the bonding jumper must be at least _____ percent of the largest phase conductor.

(a) 15   (b) 10   (c) 12 1/2   (d) 25

5. For large-capacity installations, where heavy-capacity feeders are installed to provide for future additions or changes, the rating of feeder protective devices shall be based on _____.

(a) largest motor x 1.25   (b) ampacity of feeder conductor
(c) connected load   (d) F.L.C. x demand factor

6. A 30 amp remote panelboard containing conductors protected by a 30 amp fuse shall have a _____ equipment grounding conductor minimum.

(a) #12   (b) #10   (c) #8   (d) #6

7. Electrical current is measured in terms of _____.

(a) electrical pressure   (b) electrons passing a point per second
(c) watts   (d) ohms

8. The maximum rating of a plug fuse is _____ amps.

(a) 20   (b) 30   (c) 15   (d) 40

9. A controller for a motor-compressor, serving more than one motor-compressor and other loads, shall have _____.

I. a continuous duty, F.L.C. rating
II. A locked-rotor current rating **not** less than the combined load as determined in accordance with section 440-12b

(a) I only   (b) II only   (c) either I or II   (d) both I and II

10. A multi-wire circuit may supply _____.

(a) 115/230 volt to only one utilization equipment
(b) 115/230 volt where all ungrounded conductors are opened simultaneously
(c) (a) and (b)
(d) none of these

11. _____ shall be grounded unless they are insulated from ground and from other conductor surfaces and are inaccessible to unauthorized persons.

I. Signs   II. Troughs   III. Terminal boxes

(a) I only   (b) II only   (c) III only   (d) I, II and III

12.Flexible cords shall be so connected to devices and to fittings that _____ will not be transmitted to joints or terminals.

(a) shock   (b) tension   (c) heat   (d) voltage

13. In an industrial commercial (loft) building, where the actual number of general purpose receptacle outlets are unknown, an additional load of _____ volt-amps per square foot shall be included in the load calculation.

(a) 1/2 va   (b) 1 va   (c) 1 1/2 va   (d) none of these

14. The branch circuit conductors to one or more units of a data processing system shall have an ampacity of what percent of the total connected load?

(a) 80   (b) 100   (c) 125   (d) 200

15. When a controller is **not** within sight from the motor location the disconnect shall be capable of being _____ in the open position.

(a) up   (b) down   (c) locked   (d) shut-off

16. Circuits for electrical cranes operating in Class III locations over combustible fibers shall not be
____.

(a) grounded   (b) pulled in raceways   (c) spliced   (d) THWN conductors

17. Single conductors in a cable tray shall be securely bound in circuit groups to prevent ____ due
to fault-current magnetic forces.

(a) current unbalance   **(b) inductive reactance**
(c) excessive movement   **(d) voltage surges**

18. Class II hazardous location is where ____.

(a) gases and vapors are present
(b) combustible dust is present
(c) fibers and flyings are present
(d) radioactive material is present

19. A transverse metal raceway for electrical conductors, furnishing access to predetermined cells
of a precast cellular concrete floor, which permits installation of conductors from a distribution center
to the floor cells is called ____.

(a) an underfloor raceway   **(b) a header duct**
(c) a cellular raceway   **(d) a mandrel**

20. In residential, separate outbuildings or garages, what type snap switches can be used as a
disconnecting means?

(a) 2-way and 3-way   **(b) 3-way and 4-way**
(c) 5-way and 3-way   **(d) none of these are permitted**

21. Heaters installed within ____ feet of the outlet of an air-moving device, heat pump, A/C, elbows,
baffle plates, or other obstructions in duct work may require turning vanes, pressure plates, or other
devices on the inlet side of the duct heater to assure an even distribution of air over the face of the
heater.

(a) 2   (b) 3   (c) 4   (d) 6

22. Which of the following statements about MI cable is correct?

(a) it may be used in any hazardous location
(b) a single run of cable shall not contain more than the equivalent of 4 quarter bends
(c) it shall be securely supported at intervals not exceeding 10 feet
(d) it may be mounted flush on supporting surfaces in a wet location

23. Receptacles located _____ feet above the floor are not counted in the required number of receptacles along the wall.

(a) 4   (b) 5   (c) 5 1/2   (d) none of these

24. The maximum size of a receiving station outdoor antenna conductor, where the span is 75 feet, shall be at least _____ if a copper-clad steel conductor is used.

(a) #10   (b) #12   (c) #14   (d) #17

25. The minimum common return conductor size required in an electric organ control circuit is a _____ conductor.

(a) #14   (b) #18   (c) #20   (d) #22

26. The minimum thickness of the sealing compound in Class I, Division I and II locations shall not be less than the trade size of the conduit and in no case less than _____.

(a) 3/16"   (b) 3/8"   (c) 1/2"   (d) 5/8"

27. What is the cross-sectional area of a 2" conduit?

(a) 3.36   (b) 1.34   (c) 2.016   (d) 2.067

28. Multi-outlet assembly may be used _____.

(a) where concealed   (b) in hoistways
(c) in dry locations   (d) in storage battery rooms

29. The rating of the surge arrester shall be _____ the maximum continuous phase-to-ground power frequency voltage.

I. equal to   II. less than   III. greater than

(a) I only   (b) I or II   (c) I or III   (d) II only

30. Which of the following is an acceptable wiring method for the forming shell for underground sound equipment?

I. rigid metal conduit   II. IMC brass   III. EMT   IV. rigid nonmetallic conduit

(a) I only   (b) II and III only   (c) I, II and IV   (d) IV only

31. Hoistway is a _____ in which an elevator or dumbwaiter is designed to operate.

(a) shaftway   (b) hatchway   (c) well hole   (d) all of these

32. Above ground storage tanks shall be classified _____ when the space between 5 feet and 10 feet from open end of vent extends in all directions.

(a) Class I, Division I   (b) Class I, Division II
(c) Class II, Division I   (d) Class II, Division II

33. Which of the following is required for temporary wiring?

(a) no branch-circuit conductors shall be laid on the floor
(b) all branch circuits shall orginate in an approved panelboard
(c) all conductors shall be protected as provided in Article 240
(d) all of these

34. In other than dwelling-type occupancies, each electrically heated appliance or group of appliances intended to be applied to combustible material shall be provided with a _____.

(a) light   (b) thermostat   (c) signal   (d) warning

35. The grounding conductor for secondary circuits of instrument transformers and for instrument cases shall **not** be smaller than #12 _____.

I. metal   II. aluminum   III. copper

(a) I only   (b) II only   (c) III only   (d) I, II and III

36. Fixtures or lampholders should have no live parts normally exposed to contact unless they are _____.

(a) rosette type 6 feet above the floor
(b) cleat type located at least 8 feet above the floor
(c) both (a) and (b)
(d) neither (a) nor (b)

37. For a feeder supplying household cooking equiupment and electric clothes dryers the minimum unbalanced load on the neutral conductor shall be considered as _____ percent of the load on the ungrounded conductors.

(a) 40   (b) 50   (c) 70   (d) 80

38. Where a double-throw knife switch has a vertical throw, a(n) _____ device shall be provided to hold the blades in the open position when so set.

(a) closed   (b) automatic   (c) locking   (d) tamperproof

39. Means shall be provided to disconnect the _____ of all fixed electric space heating equipment from all ungrounded conductors.

I.   heater
II.   motor
III.   controller
IV.   supplementary overcurrent protective devices

(a) I and II only   (b) II and IV only   (c) I and IV   (d) I, III and IV only

40. Where required, drawings for feeder installations must be submitted before _____.

(a) completion of installation   **(b) beginning of installation**
(c) use of feeders   **(d) all of these**

41. AC voltage of the system is not _____.

(a) EMF   (b) effective   (c) average   (d) RMS

42. How often must 1" rigid metal conduit be supported?

(a) every 10 feet   (b) every 12 feet   (c) every 14 feet   (d) every 20 feet

43. A thermal barrier shall be required if the space between the resistors and reactors and any conductor would be _____ inches.

(a) 12   (b) 15   (c) 18   (d) 24

44. Festoon lighting is a string of outdoor lights suspended between two points more than _____ feet apart.

(a) 8   (b) 10   (c) 15   (d) 25

45. _____ signs shall be marked with input amperes at full load and input voltage.

(a) All   (b) Incandescent   (c) Electric-discharge lamp   (d) Neon

46. Which of the following is not required on a motor nameplate?

(a) horsepower   **(b) manufacturer's identification**
(c) watts   **(d) voltage**

47. Fire protective signaling circuits shall be grounded, the only exception is DC power-limited circuits having a maximum current of _____ amperes.

(a) 0.300   (b) 1.5   (c) 0.030   (d) 2.34

48. FCC cable can be installed under carpet squares no larger than _____ inches.

(a) 24   (b) 30   (c) 36   (d) 48

49. Plastics machinery is defined as _____.

I. a portable machine used to shape or form plastic
II. a power-driven machine not portable by hand

(a) I only   (b) II only   (c) I and II   (d) neither I nor II

50. The vertical clearances of all service-drop conductors shall be based on conductor temperature of _____, no wind, with final unloaded sag in the wire, conductor, or cable.

(a) 60 degrees C   (b) 90 degrees C   (c) 60 degrees F   (d) 140 degrees F

51. Lead wires, furnished as a part of weatherproof lampholders and intended to be exposed after installation, shall be _____.

I.   stranded
II. rubber-covered
III. sealed in place or otherwise raintight

(a) I only   (b) II and III only   (c) I and III ony   (d) I, II and III

52. Splices and taps shall be permitted within a wireway provided they are accessible. The conductor including splices and taps shall not fill the wireway to more than _____ percent of its area at that point.

(a) 25   (b) 80   (c) 125   (d) 75

53. Escalator motors shall be classified as _____ duty.

(a) intermittent   (b) varying   (c) short-time   (d) continuous

54. Which of the following are approved wiring methods for hazardous Class II locations?

(a) mineral-insulated metal-sheathed cable   **(b) underplaster extensions**
(c) liquid-tight flexible metal conduit   **(d) bare conductor feeders**

55. The grounding conductor for a TV antenna shall not be smaller than a _____ copper.

(a) #6   (b) #8   (c) #10   (d) #12

56. Which of the following are motor starting methods listed in the N.E.C.?

I. reactor   II. resistor   III. autotransformer   IV. full-voltage

(a) II and III only   (b) I, III and IV only   (c) I, II and III only   (d) I, II, III and IV

57. What is the area of square inches for a #14 RHH with an outer covering?

(a) .0230   (b) .0327   (c) .0135   (d) .0206

58. The maximum length of a bonding jumper on the outside of a raceway is _____ feet.

(a) 3   (b) 6   (c) 25   (d) none of these

59. The minimum conductor size of flexible cords used with centralized sound distribution systems other than power supply conductors is _____.

(a) #26   (b) #22   (c) #18   (d) #16

60. What is the minimum size fixture wire?

(a) #16   (b) #18   (c) #20   (d) #22

61. Where practicable, dissimilar metals in contact anywhere in the system shall be avoided to eliminate the possibility of _____.

(a) hysteresis   (b) galvanic action   (c) specify gravity   (d) resistance

62. Types of resistive heaters are _____.

(a) heating blanket   (b) heating tape   (c) heating barrel   (d) (a) and (b)

63. Portable outdoor signs shall be protected by _____.

(a) current limiting fuses   **(b) instantaneous breakers**
(c) GFCI   **(d) time delay fuses**

64. Which of the following need not be grounded?

(a) motion picture projection equipment   **(b) electrically operated cranes**
(c) metal service raceways   **(d) electric furnace (industrial)**

65. In auxiliary gutters where the voltage on bus bars is 150 volts and the bars are opposite polarity, held free in the air, the minimum spacing between the parts is _____.

(a) 3/4"   (b) 1"   (c) 1 1/2"   (d) 2"

66. A forming shell shall be provided with a number of grounding terminals that shall be _____ the number of conduit entries.

(a) one more   (b) two more   (c) same as   (d) none of these

67. Suitable covers shall be installed on all boxes, fittings, and similar enclosures to prevent accidental contact with _____ parts or physical damage to parts or insulation. Over 600 volts nominal.

(a) energized   (b) mechanical   (c) electrical   (d) none of these

68. A 1000 watt incandescent lamp shall have a _____ base.

(a) mogul   (b) standard   (c) admedium   (d) copper

69. What kind of lighting loads does the Code state there shall be no reduction in the size of the neutral conductor?

(a) dwelling unit   (b) hospital   (c) electric-discharge   (d) motel

70. An ungrounded feeder circuit cable must be buried at least _____ inches.

(a) 24   (b) 36   (c) 12   (d) 14

OPEN BOOK
EXAM
#3

70 QUESTIONS
TIME LIMIT - 2 HOURS

TIME SPENT ☐ MINUTES

SCORE ☐ %

1. In damp or wet locations, cabinets and cutout boxes of the surface type shall be mounted so there is at least _____ air space between the enclosure and the wall or other supporting surface.

(a) 6.35"   (b) 6"   (c) 3/4"   (d) 1/4"

2. An 86 cubic inch junction box has two 1/2" rigid metal conduits entering the box. The conduits are supported within 36" of the box. The conduits enter the box through threaded hubs, one conduit through the top of the box. Select the correct statement concerning the Code requirements for this type of installation.

(a) The conduits must be supported within 18" of the box.
(b) The conduits must be supported within 12" of the box.
(c) The Code requires this box to be supported.
(d) The Code does not require this box to be supported.

3. The switch or breaker for fixed outline lighting installations must be located _____ or be capable of being locked open.

I. within sight of sign   II. within 50' of sign   III. within 75' of sign

(a) I only   (b) II only   (c) I and II only   (d) I and III only

4. Thermoplastic insulation for fixture wires may be deformed at _____ temperatures where subjected to pressure, requiring care to be exercised during installation and at points of support.

I. 0 degrees F   II. 72 degrees F   III. 212 degrees F

(a) I only   (b) II only   (c) III only   (d) II and III only

5. The installation of underfloor raceways shall be permitted _____.

I.   beneath the surface of concrete
II. beneath floor material other than concrete
III. where laid flush with the concrete floor and covered with linoleum

(a) I only   (b) II only   (c) III only   (d) I, II and III

6. FCC carpet squares that are adhered to the floor shall be attached with _____.

(a) tacking strip   (b) release-type adhesive   (c) glue   (d) none of these

7. Electric equipment with a metal enclosure or with a nonmetallic enclosure listed for the use and having adequate _____ producing characteristics, and associated wiring material suitable for the ambient temperature shall be permitted to be installed in other space used for environmental air unless prohibited in this Code.

I. fire-resistant   II. low-smoke

(a) I or II   (b) I and II   (c) I only   (d) II only

8. Conductors shall be _____ to provide ready and safe access in underground and subsurface enclosures, into which persons enter for installation and maintenance.

(a) readily accessible   (b) exposed   (c) racked   (d) enclosed

9. What is the ampacity of four #6 THW copper current-carrying conductors enclosed in schedule 80 PVC conduit, 8 foot in length entering a trench?

(a) 65 amp   (b) 52 amp   (c) 44 amp   (d) 40 amp

10. Control circuit devices with screw-type pressure terminals used with a #14 or smaller copper conductors shall be torqued to a minimum of _____ pound-inches unless identified for a different torque value.

(a) 5   (b) 6   (c) 7   (d) 8

11. A metal pole supporting a light fixture shall have an accessible handhole not less than _____ inches having a raintight cover shall provide access to the supply raceway or cable termination within the pole or pole base.

(a) 2 x 6   (b) 3 x 4   (c) 2 x 4   (d) 4 x 8

12. Liquidtight flexible metal conduit may be used in which of the following locations?

(a) in areas that are both exposed and concealed
(b) in areas where the ambient temperature is to be greater than 194 degrees C
(c) in areas that are subject to physical damage
(d) in connection areas for gasoline dispensing pumps

13. The minimum number of receptacles in a patient bed location of a hospital general care area should be _____.

(a) one   (b) two   (c) three   (d) four

14. Surge arrester grounding conductors that run in metal enclosures should be _____.

(a) bonded on one end of the enclosure only   **(b) bare**
(c) bonded at both ends of the enclosure   **(d) insulated**

15. On a recreational vehicle, the chassis rail and any welded addition thereto of metal thickness of _____ MSG or greater shall be used for the frame.

(a) 14   (b) 16   (c) 20   (d) 22

16. Heating panels installed under floor covering, shall not exceed _____ watts per square foot of heated area.

(a) 6   (b) 8   (c) 12   (d) 15

17. Where one side of the motor control circuit is grounded, the motor control circuit shall be so arranged that an _____ ground in the remote-control devices will not start the motor.

(a) intentional   (b) accidental   (c) isolated   (d) low-voltage

18. The disconnecting means for a motor shall be a _____.

I. motor-circuit switch rated in horsepower   II. circuit breaker   III. branch circuit

(a) I only   (b) II only   (c) III only   (d) I or II

19. Plate electrodes of nonferrous metal shall be at least _____ inches in thickness.

(a) 0.006   (b) 0.06   (c) 0.25   (d) 0.75

20. A discharge circuit of a capacitor shall _____.

I. be permanently connected to the terminals of the capacitor
II. be provided with automatic means of connecting it to the terminals of the capacitor bank after disconnection of the capacitor from the source of supply

(a) I only   (b) II only   (c) both I and II   (d) either I or II

21. A 120/208v, three-phase feeder supplies a continuous balanced incandescent lighting load of 270 amps. What is the minimum allowable load for the neutral conductor?

(a) 249 amps   (b) 270 amps   (c) 297 amps   (d) 338 amps

22. Which of the following may be used to insure continuity of the equipment grounding conductor in a branch circuit?

I. Connections to lampholders   II. Terminals of receptacles

(a) I only   (b) II only   (c) both I and II   (d) neither I nor II

23. Pendant fixtures shall be suspended in a Class I location by rigid non-metallic conduit if _____.

I. not more than 12" long   II. braced if over 12" long

(a) I only   (b) II only   (c) both I and II   (d) neither I nor II

24. Individual overcurrent protection for an appliance panelboard may be omitted if the panelboard feeder has overcurrent protection equal to or less than _____.

(a) the combined capacity of the overcurrent device in the panelboard
(b) 200 amp
(c) the panelboard rating
(d) the feeder ampacity

25. Power and control tray cable can be installed _____.

I. as open cables on brackets or cleats   II. in cable trays in hazardous locations

(a) I only   (b) II only   (c) both I and II   (d) neither I nor II

26. For a legally required standby system, power will be available for the application within _____.

(a) 30 seconds   (b) 10 seconds   (c) 2 minutes   (d) one minute

27. Which of the following statements about grounding conductors is true?

I. Must conduct safely any ground fault imposed on it.
II. Must have sufficiently low impedance to limit voltage to ground.

(a) I only   (b) II only   (c) both I and II   (d) neither I nor II

28. 4160v feeder, in no case shall the fuse rating in continuous amperes exceed three times, or the long-time trip element setting of a breaker _____ times, the ampacity of the conductor.

(a) 3   (b) 4   (c) 6   (d) 8

29. Conductors installed in rigid nonmetallic conduit operating at 66 kv, covered with 2" of concrete shall have a minimum cover of _____ inches.

(a) 30   (b) 18   (c) 24   (d) 42

30. The disconnecting means for portable X-ray equipment operating on 120v, 30 ampere or less branch circuits may be a _____.

I. general duty snap switch loaded not more than 80%
II. grounding type attachment plug and receptacle

(a) I only   (b) II only   (c) both I and II   (d) neither I nor II

31. A transformer case need not be grounded if _____.

I. inside a fenced enclosure   II. mounted 12' up a wood pole

(a) I only   (b) II only   (c) both I and II   (d) neither I nor II

32. An auxiliary gutter shall not extend a greater distance than _____ feet beyond the equipment which it supplements.

(a) 10   (b) 20   (c) 30   (d) 40

33. The power supply lines for portable switchboards in a theater shall be sized considering the neutral as a _____.

(a) non-current carrying conductor   (b) current-carrying conductor
(c) bonding conductor   (d) none of these

34. Bonding together of all separate grounding electrodes will limit _____ between them and between their associated wiring systems.

(a) potential differences   (b) high frequencies   (c) stray currents   (d) arcing

35. The fuse is designed so that discharged gases will not ignite or damage insulation in the path of the discharge or propagate a flashover to or between grounded members or conduction members in the path of the discharge when the distance between the vent and such insulation or conduction members conforms to manufacturer's recommendations. This is called a _____.

(a) nonvented power fuse   (b) controlled vented power fuse
(c) expulsion fuse   (d) power fuse unit

36. The voltage limitation for electrical nonmetallic tubing is _____ volts.

(a) 600   (b) 500   (c) 450   (d) 300

37. Mobile home disconnecting means shall be located not less than _____ feet above finished grade or working platform.

(a) 8   (b) 6   (c) 4   (d) 2

38. Listed baseboard heaters include instructions which may not permit their installation _____ dwelling unit receptacle outlets.

(a) above   (b) below   (c) beside   (d) adjacent

39. In commercial garages all 125v single-phase, 15 and 20 amp receptacles, where _____ are used, shall have GFCI protection for personnel.

I. portable lighting devices
II. electrical hand tools
III. electrical automotive diagnostic equipment

(a) I only   (b) II only   (c) III only   (d) I, II and III

40. Transformers and transformer vaults shall be _____ to qualified personnel for inspection and maintenance.

(a) accessible   (b) readily accessible   (c) externally operable   (d) none of these

41. Which of the following is the most acceptable grounding electrode?

(a) a driven steel approved 8' ground rod
(b) building foundation steel
(c) well casing
(d) an underground cold water metal piping system

42. The allowable distance between supports of nonmetallic-sheathed cable installed in an on site constructed, one-family dwelling is a maximum _____ feet.

(a) 2   (b) 3 1/2   (c) 4   (d) 4 1/2

43. The _____ shall not be less than the noncontinuous load plus 125% of the continuous load.

(a) branch-circuit rating   (b) continuous load   (c) non-continuous load   (d) conductor size

44. Double-pole switched lampholders supplied by the ungrounded conductors of a circuit, the switching device of lampholders of the switched type shall _____ disconnect both conductors of the circuit.

(a) separately   (b) simultaneously   (c) readily   (d) automatically

45. Any two exposed conductive surfaces in the patient vicinity shall not exceed the following potential differences at frequencies of 1000 hertz or less measured across a _____ ohm resistance.

(a) 50   (b) 100   (c) 500   (d) 1000

46. Exposed non-current carrying metal parts of fixed equipment must be grounded under all of the following conditions except _____.

(a) where supplied by metal clad wiring
(b) electrically heated devices with the frames insulated from ground
(c) where in electrical contact with metal
(d) where located in damp locations

47. A metal underfloor raceway _____.

I. must be covered if 4" wide   II. shall be covered with concrete but flush with grade

(a) I only   (b) II only   (c) both I and II   (d) neither I nor II

48. A 120v lighting circuit and a 277v lighting circuit are installed in the same raceway, the 120v circuit has a white colored grounded conductor, the 277v grounded conductor would be _____.

(a) white   (b) gray   (c) white with a green stripe   (d) white with a yellow stripe

49. A totally enclosed positive pressure ventilating motor is to be used as a power source in a Class I, Division I area. Which of the following is correct?

I. The time delay interlock must be arranged to allow the thorough purging of the motor enclosure before energizing of the motor.
II. Must be arranged to automatically de-energize the motor when the air supply fails.

(a) I only   (b) II only   (c) both I and II   (d) neither I nor II

50. Signs and outline system enclosures shall have not less than the vertical and horizontal clearances from open conductors specified in Article _____.

(a) 230   (b) 410   (c) 600   (d) 225

51. A metallic water system is used as a grounding means. Which of the following statements is/are true?

I. must be bonded around meter   II. must bond around insulated joints

(a) I only   (b) II only   (c) both I and II   (d) neither I nor II

52. Conductors installed outdoors at 7200v, would have a minimum clearance phase to phase of ____.

(a) 4 1/2"   (b) 4"   (c) 7"   (d) 5 1/2"

53. Portable tools or appliances are not intended to be used in damp, wet or conductive locations unless they are ____.

I. grounded   II. double insulated   III. supplied through an isolating transformer

(a) I only   (b) II only   (c) III only   (d) I, II or III

54. Power-limited fire protective signaling circuits where installed in exposed cable, located within 7 feet of the floor, cable shall be securely fastened in an approved manner at intervals of not more than ____ inches.

(a) 18   (b) 24   (c) 30   (d) 36

55. Conductors operating at less than 50 volts and not covered in Articles 650, 725 and 760, shall not be smaller than ____ copper or equivalent.

(a) #16   (b) #18   (c) #14   (d) #12

56. Armored cable installed in thermal insulation shall have conductors rated at ____. The ampacity of the cable installed in these applications shall be that of 60 degree C conductors.

(a) 60 degrees C   (b) 194 degrees F   (c) 75 degrees C   (d) 90 degrees F

57. Indoor antennas and indoor lead-ins shall be permitted to occupy the same box or enclosure with conductors of other wiring systems where separated from such other conductors by an effective permanently installed ____.

(a) wall   (b) divider   (c) insulator   (d) barrier

58. Components of lighting track systems of different voltages shall not be ____.

(a) connected   (b) interchangeable   (c) polarized   (d) none of these

59. Where separate services supply a building and are required to be connected to a grounding electrode, _____ shall be used.

(a) multiple electrodes   (b) the same grounding electrode
(c) two electrodes   (d) paralleled electrodes

60. Snap switches can be grouped or ganged in outlet boxes if voltages between adjacent switches do not exceed _____ volts.

(a) 150   (b) 300   (c) 350   (d) 400

61. Excluding fuses and exceptions, how many overcurrent protection devices, such as trip coils, relays or thermal cutouts, are required on a three-phase motor?

(a) 0   (b) 1   (c) 2   (d) 3

62. All of the following sizes of solid aluminum conductors must be made of aluminum alloy except

_____.

(a) #14 AWG   (b) #12 AWG   (c) #10 AWG   (d) #8 AWG

63. Lighting fixtures installed over vehicle lanes inside a commercial garage shall be installed a minimum of _____ feet.

(a) 8   (b) 10   (c) 12   (d) 15

64. The bare neutral of aluminum conductors may not be used underground on a service except _____.

(a) if protected at not more than 20a   (b) if protected with oxide inhibiter
(c) if installed in aluminum conduit   (d) if arranged in a cable assembly

65. Which of the following statements about circuit breakers used to provide load protection for a high voltage motor circuit is correct?

I. The breakers shall simultaneously disconnect all ungrounded conductors to the motor.
II. The breakers may sense a fault current by means of integral external sensing elements.

(a) I only   (b) II only   (c) both I and II   (d) neither I nor II

66. The maximum allowable rating for overcurrent devices for an AC transformer welder is _____ percent of the rated primary current.

(a) 100   (b) 125   (c) 150   (d) 200

67. In a Class I, Division II area, bonding can be accomplished by _____.

I. double locknuts   II. bonding wire pulled in conduit

(a) I only   (b) II only   (c) both I and II   (d) neither I nor II

68. In a grounded system the grounding electrode conductor shall be connected to the following _____.

I. at any accessible point from the load end of the service drop
II. including the terminal to which the grounded service conductor is connected at the service
   disconnecting means

(a) I only   (b) II only   (c) both I and II   (d) neither I nor II

69. Two conductors in a conduit, what is the percent of allowable fill?

(a) 31%   (b) 40%   (c) 53%   (d) 80%

70. An accessible means external to enclosures for connecting intersystem bonding and grounding
conductors shall be provided at the service by at least one of the following means _____.

I. exposed grounding electrode conductor
II. exposed metallic service raceways
III. approved means for the external connection of a bonding, or grounding conductor to the service
   raceway or equipment

(a) I only   (b) II only   (c) III only   (d) I, II or III

OPEN BOOK EXAM #4

70 QUESTIONS
TIME LIMIT - 2 HOURS

TIME SPENT ☐ MINUTES

SCORE ☐ %

1. Where NM cable is used with boxes no larger than a nominal size 2 1/4" x 4" mounted in walls and where the cable is fastened within _____ inches of the box measured along the sheath and where the sheath extends into the box no less than 1/4", securing the cable to the box shall not be required.

(a) 8   (b) 10   (c) 12   (d) 24

2. A bare copper conductor can be used in an underground service _____.

(a) where judged suitable for soil conditions
(b) where used in a raceway
(c) without regard to soil conditions where part of cable assembly identified for underground
    use
(d) all of the above

3. All of the following may be used as grounding electrode connections for fire alarm systems except
_____.

(a) service equipment enclosures
(b) continuous extensive underground gas piping
(c) hot water piping
(d) metallic power-service conduit

4. "Direct grade level access" is defined as being located not more than _____ above grade level.

(a) 6 1/2'   (b) 6' 3"   (c) 6' 4"   (d) 7'

5. A main bonding jumper shall be a _____.

I. wire   II. screw   III. bus

(a) I only   (b) I or II only   (c) I or III only   (d) I, II or III

6. The earth shall not be used as the sole _____ conductor.

(a) equipment grounding   (b) grounded   (c) neutral   (d) bonding

7. Each receptacle of DC plugging boxes shall be rated at not less than _____ amps.

(a) 15   (b) 20   (c) 25   (d) 30

8. A protective layer which is installed between the floor and type FCC flat conductor cable to protect the cable from physical damage and may or may not be incorporated as an integral part of the cable is the _____.

(a) transition assembly   (b) outer sheath   (c) bottom shield   (d) header

9. An area must be classed as a Class II hazardous location if it contains _____.

(a) combustible dust   (b) ignitable vapors   (c) flammable gases   (d) ignitable fibers

10. Illumination is mandatory for service equipment or panelboards in a dwelling unit if the service to the unit exceeds _____ amps.

(a) 100   (b) 200   (c) 300   (d) 400

11. Which of the following statements about lightning protection is (are) true?

I. Lightning rods should be spaced at least 6 feet away from noncurrent-carrying metal parts of electrical equipment.
II. If lightning protection is required for an irrigation machine, a driven ground rod should be connected to the machine at a stationary point.

(a) I only   (b) II only   (c) both I and II   (d) neither I nor II

12. If the appliance is provided with a single-pole switching device, the attachment plug shall be _____.

I. of the grounding type   II. polarized

(a) I only   (b) II only   (c) I or II   (d) neither I nor II

13. Listed equipment protected by a system of double insulation, or its equivalent, shall not be required to be grounded. Where such a system is employed, the equipment shall be _____.

(a) labeled   (b) approved   (c) distinctively marked   (d) identified

14. The ampacity of a #12 THW conductor used with a 15-minute motor on a monorail hoist installed in a raceway containing a total of four current-carrying conductors would be _____ amps.

(a) 20   (b) 30   (c) 33   (d) 36.96

15. No premises wiring, with a grounded conductor, shall be electrically connected to a supply system unless the supply system contains _____.

(a) a grounded conductor   **(b) a wiring design**
(c) protection   **(d) none of these**

16. The prime mover of an emergency generator set _____.

I.   must be provided with an automatic means for starting
II. must be provided with an automatic means of transferring from one fuel supply to another, where dual fuel supplies are used
III. must have on-site fuel supply sufficient to operate the prime mover at full demand for 2 hours

(a) I only   (b) II only   (c) III only   (d) I, II and III

17. Open core-and-coil type transformers shall be limited to _____ applications.

(a) outline lighting installations   **(b) outdoor portable signs**
(c) fixed signs   **(d) indoor portable**

18. A cord connector that is supported by a permanently installed cord pendant shall be considered a(an) _____.

(a) receptacle outlet   (b) permanent cord   (c) lighting outlet   (d) outlet device

19. Auxiliary equipment for electric-discharge lamps shall be enclosed in noncombustible cases and _____.

I.   at least 3 inches away from any combustible material
II. shall not be over 1500 watts
III. treated as a source of heat

(a) I only   (b) II only   (c) III only   (d) I, II and III

20. Stage cables used in motion picture studios for stage lighting shall be protected by means of overcurrent devices set at not more than _____ of the values given in the appropriate Code table.

(a) 500%   (b) 400%   (c) 200%   (d) 100%

21. Amplifier output circuits carrying audio-program signals of 70 volts or less whose open-circuit voltage will not exceed _____ volts shall be permitted to employ Class 2 or Class 3 wiring as covered in Article 725.

(a) 100   (b) 120   (c) 150   (d) 300

22. ____ of conductors in rigid nonmetallic conduit shall be made only in junction, outlet boxes or conduit bodies.

(a) Splices   (b) Splices and taps   (c) Connections   (d) none of these

23. All devices provided with terminals for the attachment of conductors and intended for connection to more than one side of the circuit shall have the ____ identified.

(a) conductors   (b) terminals   (c) sides   (d) none of these

24. ____ of insulating material shall be permitted to be used without boxes in exposed cable wiring.

I. Switch devices   II. Outlet devices   III. Tap devices

(a) I only   (b) II only   (c) III only   (d) I, II and III

25. A nursing home is a building or part thereof used for the lodging, boarding and nursing care, on a 24-hour basis, of ____ or more persons.

(a) 4   (b) 12   (c) 50   (d) 100

26. Which of the following must be provided with GFCI?

(a) dishwashers   (b) fountains   (c) outdoor lights   (d) refrigerators

27. A resistance welder shall have overcurrent primary protection set at not more than ____ percent.

(a) 200   (b) 300   (c) 250   (d) 125

28. The minimum radius of a bend in a 1" MI cable is ____ inches.

(a) 4   (b) 5   (c) 10   (d) 12

29. Service conductors on a 4600v system shall not be smaller than ____ unless in cable.

(a) #6   (b) #4   (c) #2   (d) #1

30. In cellular metal floor raceways all of the following are true except ____.

(a) splices and taps can be made in junction boxes
(b) disconnected outlets are removed
(c) entry boxes are installed flush to the floor
(d) the combined cross sectional fill cannot exceed 45%

31. Non-shielded high-voltage cables shall be installed in _____ conduit encased in not less than 3" of concrete.

I. rigid PVC   II. IMC   III. rigid metal

(a) I only   (b) II only   (c) III only   (d) I, II or III

32. On solar photovoltaic system; Ampacity of conductors and overcurrent devices shall not be less than _____ percent of the computed current.

(a) 150   (b) 100   (c) 125   (d) 200

33. A run of flexible metal conduit may be used as an equipment grounding conductor if the conductors are protected at _____.

(a) 20a or more   (b) 20a or less   (c) 30a or more   (d) 30a or less

34. Which of the following would be permissible in a panelboard for general lighting?

(a) 4 - 3 pole breakers, 12 - 2 pole breakers, 8 - 1 pole breakers
(b) 6 - 3 pole breakers, 10 - 2 pole breakers, 8 - 1 pole breakers
(c) 5 - 3 pole breakers, 11 - 2 pole breakers, 5 - 1 pole breakers
(d) 5 - 3 pole breakers, 8 - 2 pole breakers, 12 - 1 pole breakers

35. The service to a mobile home may consist of _____.

I. 3 - 50a separate services   II. one permanently connected feeder

(a) I only   (b) II only   (c) both I and II   (d) neither I nor II

36. A motor is connected to the electrical system by #10 THW conductors that are protected by a double pole 40 amp breaker. These conductors are enclosed in a 1/2" liquidtight flexible metal conduit that is 4 foot long and is terminated in connectors that are approved for grounding. Which of the following is the correct statement concerning this installation?

(a) The 1/2" liquidtight conduit will not permit the installation of two #10 THW's due to the 40% fill requirement.
(b) The conduit must contain a grounding conductor since it is more than 3' in length.
(c) The conduit must contain a grounding conductor since the conductors it contains are fused at more than 20 amps.
(d) The conduit must contain a grounding conductor since it contains conductors that are connected to a motor load.

37. You are installing a 75 foot run of 2" rigid metal conduit, using threaded couplings. The Code requires you to support this conduit within 3 feet of termination, and then a maximum of _____ feet apart.

(a) 3   (b) 8   (c) 12   (d) 16

38. The #4 solid copper conductor connecting the grounding electrode to the service has been accidentally cut. Which of the following would be acceptable connectors for splicing this conductor?

I.   a split bolt connector
II. a compression sleeve type connector
III. a wire nut

(a) I only   (b) I or II only   (c) I, II or III   (d) none of these are acceptable

39. In panelboards, where the voltage on busbars is 150 volts and the bars are opposite polarity, held free in air, the minimum spacing between the parts is _____.

(a) 3/4"   (b) 1"   (c) 1 1/2"   (d) 2"

40. _____ is a combination consisting of a compressor and motor, both of which are enclosed in the same housing, with no external shaft or shaft seals, the motor operating in the refrigerant.

I.   Motor-compressor
II. Hermetic refrigerant motor-compressor
III. Air-conditioning equipment

(a) I only   (b) II only   (c) III only   (d) I, II and III

41. A 3" conduit, not more than _____ inches in length, connecting a gutter and a switch case, may have a conductor fill of 60% of the internal cross-sectional area.

(a) 6   (b) 12   (c) 18   (d) 24

42. The service entrance conductors are #350 MCM copper. The minimum size grounding electrode conductor allowed by the Code would be a _____ copper, if it were the sole connection to a made electrode.

(a) #4   (b) #2   (c) #6   (d) #8

43. Supplementary overcurrent devices shall not be required to be _____.

(a) accessible   (b) readily accessible   (c) continuous duty   (d) adjustable

44. Where single conductors are installed in a triangular configuration in uncovered cable trays, with a maintained space of not less than _____ times one conductor diameter between circuits, the ampacity of #1/0 AWG and larger cables shall not exceed the allowable ampacities of two or three single insulated conductors rated 0 through 2000 volts supported on a messenger.

(a) 3   (b) 2   (c) 2.15   (d) 2.75

45. All 125v, single-phase 15 and (or) 20 amp receptacles installed _____ shall have GFCI protection for personnel.

I.   in boathouses
II. within 6' of the kitchen sink above counter top surfaces
III. in a basement

(a) I and II only   (b) II and III only   (c) II only   (d) I, II and III

46. All electric spa or hot tub water heaters shall be listed and have the heating elements subdivided into loads not exceeding _____ amperes and protected at not more than _____ amperes.

(a) 45 ... 50   (b) 48 ... 60   (c) 40 ... 45   (d) 55 ... 60

47. Capacitors shall be permitted to be protected _____.

I. in groups   II. individually

(a) I only   (b) II only   (c) I or II   (d) neither I nor II

48. Listed ceiling fans that do not exceed _____ pounds in weight, with or without accessories, shall be permitted to be supported by outlet boxes.

(a) 35   (b) 40   (c) 45   (d) 50

49. When service entrance phase conductors are larger than 1100 MCM copper, the bonding jumper shall have an area not less than what percent of the area of the largest phase conductor?

(a) 6%   (b) 10%   (c) 12 1/2%   (d) 15%

50. Surface-mounted fluorescent lighting fixtures that contain a ballast and are to be installed on combustible, low-density cellulose fiberboard should be spaced not less than _____ inches from the surface of the fiberboard.

(a) 1/4"   (b) 1/2"   (c) 1"   (d) 1 1/2"

51. Except for one and two-family dwellings a 15 or 20 amp rated receptacle outlet shall be installed at an accessible location for the servicing of heating, A/C and refrigeration equipment. The receptacle outlet shall be located on the same roof level and within _____ feet of the equipment.

(a) 25   (b) 50   (c) 75   (d) 100

52. Where multicircuit track is installed, the load requirement of this section shall be considered to be _____ between the circuits.

(a) 180va   (b) twice the smallest load   (c) divided equally   (d) continuous

53. Select the correct statement that pertains to a general purpose 15 amp outlet that is installed in a marina.

(a) It is in violation of the Code because such outlets shall be rated not less than 20a.
(b) It shall not be located less than 5' measured horizontally from the water line.
(c) It shall be mounted so that waves will not create a problem.
(d) It is required by the Code to be of a locking and grounding type.

54. Articles _____ through _____ cover occupancies or parts of occupancies that are or may be hazardous because of atmospheric concentrations of flammable liquids, gases, or vapors, or because of deposits or accumulations of materials that may be readily ignitible.

(a) 517 through 520   (b) 511 through 517   (c) 514 through 517   (d) 514 through 600

55. Isolating switches over 600v shall be provided with a means for readily connecting the load side conductors to ground when disconnected from the _____.

(a) current   (b) equipment   (c) service cable   (d) source of supply

56. The rating of the surge arrestor shall be _____ the maximum continuous phase-to-ground power frequency voltage available at the point of application.

I. equal to   II. less than   III. greater than

(a) I only   (b) I or II only   (c) I or III only   (d) II only

57. Omission of overcurrent protection shall be permitted at points where busways are reduced in size, provided that the smaller busway doesn't extend more than _____ feet.

(a) 10   (b) 20   (c) 50   (d) 70

58. A cellular concrete floor raceway's grounding conductor shall connect the insert receptacles to a positive ground connection provided on the _____.

(a) junction box   (b) cell   (c) fitting   (d) header

59. A circuit breaker may be adjusted for _____.

I. various values of current   II. various values of time

(a) I only   (b) II only   (c) both I and II   (d) neither I nor II

60. All of the following motors are permitted in a Class III, Division I area except _____.

(a) pipe ventilated   (b) non-ventilated   (c) fan cooled   (d) water cooled

61. The maximum overcurrent device on a branch circuit supplying a ASME rated boiler is _____ amp.

(a) 40   (b) 60   (c) 100   (d) 150

62. A device supplying running overload protection may be shunted during starting a motor when it is started _____.

I. manually   II. automatically

(a) I only   (b) II only   (c) both I and II   (d) neither I nor II

63. For outdoor installations, open conductors 4160v shall have a minimum clearance of _____ phase to phase.

(a) 4.5"   (b) 5.5"   (c) 7"   (d) 12"

64. Which of the following statements is/are true?

I. On a grounded service, the grounded service neutral shall not be smaller than the grounding electrode conductor.
II. If a 1000v or less system is grounded, the grounded conductor must be run to each service.

(a) I only   (b) II only   (c) both I and II   (d) neither I nor II

65. Maximum voltage between conductors serving a submersible pump in a fountain is _____ volts.

(a) 150   (b) 250   (c) 300   (d) 600

66. Which of the following about an aircraft hangar is true?

I. Any area below the floor level shall be considered a Class I, Division I location up to the floor level.
II. The area within 5' horizontally of aircraft power plants or fuel tanks shall be considered a Class I, Division II location extending from the floor to a level 5' above the upper surface of the wings and engine housing.

(a) I only   (b) II only   (c) both I and II   (d) neither I nor II

67. All of the following are true about transformer vaults except _____.

(a) ventilated openings are calculated at 3 sq.in. per kva
(b) each doorway shall be fully louvered for ventilation
(c) door sills and curbs shall not be less than 4" in height
(d) materials shall not be stored inside the vaults

68. The _____ and the bridge frame shall be considered as electrically grounded through the bridge and trolley wheels and its respective tracks unless local conditions, such as paint or other insulating material, prevent reliable metal-to-metal contact.

(a) trolley frame   (b) track   (c) trolley wheels   (d) none of these

69. A Class I, Division I motor shall _____.

I. have a time delay start for ventilation   II. shall stop if ventilation fan fails

(a) I only   (b) II only   (c) both I and II   (d) neither I nor II

70. A buried cable, 4160v, requires a minimum depth of _____ inches.

(a) 24   (b) 30   (c) 36   (d) 42

OPEN BOOK
EXAM
#5

70 QUESTIONS
TIME LIMIT - 2 HOURS

TIME SPENT ☐ MINUTES

SCORE ☐ %

1. Fixtures shall be securely fastened to ceiling framing member by mechanical means such as _____.

I. rivets   II. screws   III. bolts

(a) II only   (b) III only   (c) II and III only   (d) I, II and III

2. Insulated conductors used in wet locations shall be _____.

(a) lead covered   (b) asbestos   (c) THHN   (d) varnished cambric

3. In dwelling units, a multiwire branch circuit supplying a split-wired receptacle is required to _____.

(a) have color-coded conductors to aid in identification
(b) carry an individual neutral for each ungrounded conductor
(c) have a provision at the source to simultaneously disconnect all
**   ungrounded conductors**
(d) have a minimum of two neutrals

4. The word transformer is intended to mean a _____ transformer, single or polyphase, identified by a single nameplate, unless otherwise indicated.

(a) group   (b) two   (c) individual   (d) step-down

5. Unused openings in boxes, raceways, and other enclosures shall be _____.

(a) closed with a device listed for such service with the equipment
(b) effectively closed
(c) closed to afford protection equivalent to the equipment wall
(d) open

6. Spacing between conduits, tubing, or raceways shall be _____.

(a) 1/4"   (b) insulated   (c) 1"   (d) maintained

7. Feeders to floating dwellings must be enclosed within _____ conduit in order to withstand the forces exerted by waves and tides.

(a) rigid metal   (b) rigid PVC   (c) liquidtight flexible   (d) EMT

8. Running open wiring on insulators, MI or MC cable, messenger-supported wiring, conductors in raceway, and other approved means on the outdoor building surfaces is permitted for circuits operating at a maximum of _____ volts nominal.

(a) 600   (b) 750   (c) 1000   (d) 4160

9. Where a single AC conductor carrying current passes through metal with magnetic properties, the inductive effect shall be minimized by _____.

I. cutting slots in the metal between the individual holes through which individual conductors pass
II. passing all the conductors in the circuit through an insulating wall sufficiently large for all the
   conductors of the circuit

(a) I only   (b) II only   (c) both I and II   (d) neither I nor II

10. For nondwelling units, it is permitted to use a _____ demand factor for that portion of a receptacle load that exceeds 10 kva.

(a) 70%   (b) 80%   (c) 50%   (d) 40%

11. Wiring over navigable water must be approved by the _____.

(a) corps of engineers   (b) U.S. Coast Guard
(c) authority having jurisdiction   (d) Department U.S. Navy

12. The phase current in a grounding autotransformer is _____ the neutral current.

(a) twice   (b) 1/2   (c) 1/3   (d) the same as

13. In a closed-loop and programmed power distribution system, the outlets shall _____.

(a) energize only when the load allows the completion of a loop
(b) be energized only when plugged-in equipment is identified
(c) be constantly energized, as is done in existing 120/240v power systems
(d) none of these

14. Electrical ducts shall include any of the electrical conduits recognized in Chapter 3 as suitable for use _____.

(a) over 600v   (b) as bus bars   (c) underground   (d) exposed

15. All disconnect means required by the Code, and each service, feeder, and branch circuit shall be suitably marked ____.

(a) with a sign "Danger of Electrocution"   **(b) "Disconnect"**
(c) and provided with a lockout means   **(d) to indicate its purpose**

16. For cables that have elliptical cross section, the cross-sectional area calculation shall be based on using ____ of the ellipse as a circle diameter.

(a) half   (b) the radius   (c) the major diameter   (d) the circumference

17. The alternate source for emergency systems ____ be required to have ground-fault protection of equipment.

I. shall   II. shall not

(a) I only   (b) II only   (c) either I or II   (d) neither I nor II

18. Where two or more single-phase ranges are supplied by a 3-phase, 4-wire feeder, the total load shall be computed on the basis of ____ maximum number connected between any two phases.

(a) twice the   (b) three times the   (c) half the   (d) none of these

19. Type ____ is a single or multiconductor solid dielectric insulated cable rated 2001 volts or higher.

(a) MI   (b) NM   (c) MC   (d) MV

20. The secondary circuits of wound-rotor alternating current motors, including ____ shall be permitted to be protected against overload by the motor-overload device.

(a) resistors   (b) controllers   (c) conductors   (d) all of these

21. Where secondary ties are used, an overcurrent device rated or set at not more than ____ percent of the rated secondary current of the transformers shall be provided in the secondary connections of each transformer.

(a) 100   (b) 150   (c) 250   (d) 300

22. Conductors supplying outlets for arc or Xenon projectors of the professional type shall not be smaller than ____.

(a) #12   (b) #10   (c) #8   (d) #6

23. A weatherproof receptacle and attachment plug having one pole for grounding shall be provided for each individual _____.

(a) letter   (b) fixture   (c) sign   (d) all of these

24. Receptacles rated ____ amperes or less directly connected to aluminum conductors shall be marked CO/ALR.

(a) 20   (b) 25   (c) 30   (d) 50

25. Each commercial building and each commercial occupancy accessible to pedestrians shall be provided at an accessible location outside the entrance, with at least one ____ for sign or outline lighting use.

(a) outlet   (b) duplex   (c) GFCI   (d) none required

26. Transformers for outline lighting installations shall have secondary current ratings not more than ____ milliamperes.

(a) 20   (b) 30   (c) 40   (d) 50

27. Electric-discharge lighting fixtures having exposed ____ shall be so installed that these parts will not be in contact with combustible material.

I. live parts   II. ballasts or transformers   III. auxiliary equipment

(a) I only   (b) II only   (c) III only   (d) I, II and III

28. A ____ circuit is a circuit in which any spark or thermal effect is incapable of causing ignition of a mixture of flammable or combustible material in air under prescribed test conditions.

(a) Low voltage   (b) Instrinsically safe   (c) Hazard-proof   (d) Explosive-proof

29. Only wiring methods recognized as ____ are included in the Code.

(a) approved   (b) suitable   (c) listed   (d) identified

30. The ampacities for concealed knob and tube wiring can be found in ____.

(a) Table 310-16   (b) Table 310-19   (c) Table 310-17   (d) Section 310-15

31. The N.E.C. covers _____.

I. installations of electrical conductors and equipment within or on public and private buildings or other structures, including mobile homes, recreational vehicles, and floating dwelling units; and other premises, such as yards, carnivals, parking and other lots, and industrial substations
II. installations of conductors that connect to the supply of electricity
III. installations of optic fiber cable

(a) I only (b) II only (c) III only (d) I, II and III

32. Circuits for _____ shall **not** be connected to any system containing trolly wires with a ground return.

(a) kitchens and laundry rooms   (b) carhouses and power houses
(c) railway stations   (d) lighting and power

33. When the kind of motor is single-phase AC or DC, and its supply system is 2-wire, single-phase AC or DC, one conductor grounded, the minimum number and the location of overload units, such as trip coils, relays or thermal cutouts shall be _____.

(a) two, one per phase in hot conductors   (b) one in the ungrounded conductor
(c) one in each conductor   (d) two, one in each phase

34. In a 6-pole machine, 360 electrical degrees is equal to _____ mechanical degrees.

(a) 60 (b) 90 (c) 120 (d) 180

35. The proper use of suitable ground detectors on ungrounded systems can provide _____.

(a) no additional protection   (b) additional protection
(c) a grounded system   (d) additional circuits

36. No _____ other than those specified as required for emergency use, shall be supplied by emergency lighting circuits.

I. appliances II. lamps III. fittings

(a) I only (b) I and II only (c) II and III only (d) I, II and III

37. The approximate area square inch of a #2 THW aluminum conductor is _____.

(a) .1473 (b) .1194 (c) .1182 (d) .1017

38. An appliance (not motor driven) is rated 1200 watts at 120 volts, with no marked nameplate, the branch circuit overcurrent device shall not exceed _____ amps.

(a) 15 (b) 20 (c) 30 (d) 40

39. A storage battery of suitable rating and capacity to supply and maintain at not less than _____ of system voltage the total load of the circuits supplying legally required standby power for a period of at least 1 1/2 hours.

(a) 100% (b) 75% (c) 50% (d) 87 1/2%

40. What is the maximum time of delay permitted for the GFI to operate where the ground-fault current is 4000 amperes?

(a) 1/2 second (b) 1 second (c) 3 seconds (d) 100 mili-seconds

41. Which of the following methods is **not** approved for conductor supports?

(a) **deflecting of cables in junction boxes**   (b) **insertion of boxes**
(c) **clamping devices**   (d) **loop connectors**

42. A separate branch circuit shall supply the _____ for each elevator.

(a) fan (b) car lights (c) emergency phone (d) emergency exit

43. An underground service installed in PVC and having a 3" concrete envelope shall be buried a minimum of _____ inches.

(a) 6 (b) 12 (c) 18 (d) 24

44. The ground fault protection system shall be tested when it is _____.

(a) installed (b) energized for the first time (c) inspected (d) manufactured

45. Where a motor is operating and live parts of the motor controller have over 150 volts to ground and might be exposed to repairmen, what must be done for its safe maintenance?

(a) two technicians must be present when servicing the motor
(b) insulating mats shall be provided
(c) a sign saying "Danger" must be installed near the motor
(d) power disconnected when working on it

46. A junction box used in a system rated 1000 volts shall have a marking on the box of _____.

(a) Caution   (b) Danger   (c) Do Not Open   (d) Danger High-Voltage Keep Out

47. In a pediatric location a tamperproof receptacle is one which by its construction is a receptacle which _____.

(a) is of the lock-type   (b) has a control switch internally
(c) cannot be opened   (d) limits access to its energized contacts

48. According to the N.E.C., high-voltage service-entrance conductors are protected by a circuit breaker if it has _____ the ampacity of the conductor for its trip setting. (Short circuit protection).

(a) 3 times   (b) 5 times   (c) 6 times   (d) 8 times

49. Which of the following locations is **not** permitted for the use of surface raceways?

(a) dry location   (b) underplaster   (c) data processing   (d) hazardous

50. In a Class I, Division II location a conduit passing through into a nonhazardous location, the sealing fitting shall be permitted _____.

(a) no seal required under the conditions   (b) on either side of the boundary
(c) on both sides of the boundary   (d) at the first fitting

51. Underfloor raceways not more than 102 millimeters wide shall have no less than _____ concrete or wood above them.

(a) 1/2 inch   (b) 3/4 inch   (c) 2 inches   (d) 4 inches

52. A conductor at a terminal which leaves a cabinet, the minimum wire bending space at the terminal shall be how much if the conductor is a #250 MCM and leaves through wall opposite its removeable terminal?

(a) 4"   (b) 4 1/2"   (c) 5"   (d) 6 1/2"

53. A community antenna system grounding conductor shall not be smaller than _____ AWG, and shall have a current-carrying capacity approximately equal to that of the outer conductor of the coaxial cable.

(a) #18   (b) #16   (c) #14   (d) #12

54. Equipment installed and likely to become energized shall be grounded at which of the following distances?

(a) 8 feet horizontal and 5 feet vertical of grounded metal objects
(b) 5 feet vertical and 5 feet horizontal of grounded metal objects
(c) 8 feet vertical and 5 feet horizontal of grounded metal objects
(d) 8 feet vertical and 8 feet horizontal of grounded metal objects

55. Boxes having an approved system of organic coatings and are installed out of doors shall be marked _____.

(a) weatherproof   (b) raintight   (c) watertight   (d) outdoor usage

56. A receptacle installed outdoors shall be considered protected from the weather by which of the following methods?

I. locating under a roof, canopies, marquees
II. covering with a snap cover
III. installing a metal hood cover over the top of the receptacle

(a) I only   (b) II and III only   (c) I and II only   (d) I, II and III

57. A cutout box installed in a wet location shall be _____.

(a) raintight   (b) weatherproof   (c) waterproof   (d) rainproof

58. Electrical nonmetallic tubing is permitted _____.

I.   concealed in walls, floors and ceilings with a 15 minute fire rating
II. embedded in concrete provided with approved fitting
III. directly buried
IV. above a suspended ceiling with a 15 minute fire rating

(a) I only   (b) I, II and IV   (c) I, II and III   (d) all of the above

59. Where load is connected to the tie at any point between transformer supply points and overcurrent protection is not provided in accordance with Article 240, the rated ampacity of the tie shall not be less than ___ percent of the rated secondary current of the largest transformer connected to the secondary tie system.

(a) 100   (b) 150   (c) 250   (d) 300

60. The branch circuit rating of a non-motor operated appliance that is continuously loaded shall have a minimum rating of _____ percent of the marked rating.

(a) 50   (b) 80   (c) 125   (d) 150

61. Each service disconnecting means shall _____ disconnect all ungrounded service conductors from the premises wiring system.

(a) automatically   (b) independently   (c) simultaneously   (d) separately

62. A conductor at a terminal which leaves a cabinet, the minimum wire bending space at the terminal shall be _____ inches if the conductor size is #2 and it does not enter the enclosure through the wall opposite.

(a) 2   (b) 2 1/2   (c) 3 1/2   (d) 4

63. A 50 volt generator which is driven by a single motor is protected by the overcurrent protecting the motor only when the generator is delivering no more than _____ percent of its full load rated current.

(a) 80   (b) 100   (c) 125   (d) 150

64. MI cable shall be permitted _____.

I.   as feeders and branch circuits
II. for wet and dry locations
III. for concealed or exposed

(a) I only   (b) II only   (c) III only   (d) I, II and III

65. If SE or USE cable consists of two or more conductors, one shall be permitted to be _____.

(a) insulated   (b) green   (c) tagged   (d) uninsulated

66. For small motors not covered by the motor tables, the locked-rotor current shall be assumed to be _____ times the full load current.

(a) three
(b) four
(c) five
(d) six

67. Fuses and circuit breakers shall be so located or _____ that persons will not be burned or otherwise injured by their operation.

(a) concealed   (b) guarded   (c) shielded   (d) elevated

68. Surface marking of conductors and cables shall be durably marked on the surface at intervals not exceeding _____ inches.

(a) 6   (b) 12   (c) 18   (d) 24

69. No cord on a portable sign shall be less than _____ feet from the ground level directly underneath.

(a) 6   (b) 7   (c) 8   (d) 10

70. The horsepower rating of the disconnecting means of hermetic refrigerant motor-compressors shall be determined from the _____.

I.   nameplate rated-load current
II. branch-circuit selection current
III. loads

(a) I only   (b) II only   (c) III only   (d) I or II only

OPEN BOOK EXAM
#6

70 QUESTIONS
TIME LIMIT - 2 HOURS

TIME SPENT ☐ MINUTES

SCORE ☐ %

1. The equipment grounding conductor in type NM cable for 15, 20 and 30 ampere branch circuits _____.

(a) may be at least one size smaller than the insulated circuit conductor
(b) must be the same size as the insulated circuit conductors
(c) is required only with aluminum cable
(d) none of these

2. Thermoplastic insulation may stiffen at temperatures colder than minus _____ degrees C requiring care be exercised during installation at such temperatures.

(a) 5   (b) 10   (c) 15   (d) 30

3. Service heads for service conductors shall be _____.

(a) raintight   (b) weatherproof   (c) rainproof   (d) watertight

4. Rigid conduit buried in an area subject to heavy vehicular traffic shall have a minimum cover of _____ inches.

(a) 6   (b) 12   (c) 18   (d) 24

5. Locations where flammable paints are dried, but in which the ventilating equipment is interlocked with the electrical equipment may be designated as _____.

(a) Class I, Division II   **(b) nonhazardous**
(c) Class II, Division II   **(d) Class II, Division I**

6. At what angle does a header attach to a floor duct?

(a) parallel   (b) straight   (c) right angle   (d) none of these

7. In Class I, Division I locations, the Code requires conduit seals adjacent to boxes containing splices if the conduit is equal to or larger than _____.

(a) 3/4"   (b) 1 1/2"   (c) 1"   (d) 2"

8. Every electric sign of any type, fixed or portable, shall be _____.

(a) listed   (b) approved   (c) permanently wired   (d) electrically isolated

9. In motion-picture studios, feeder conductors to the stage may be protected, with respect to ampacity, at a maximum value of _____ percent.

(a) 200   (b) 250   (c) 400   (d) 500

10. Service conductors run above the top level of a window shall be _____.

(a) 3 feet above the window   **(b) considered out of reach**
(c) accessible   **(d) 8 feet above window**

11. Cells in jars of conductive material shall be installed in trays of nonconductive material with not more than _____ cells in the series circuit.

(a) 16   (b) 18   (c) 20   (d) 24

12. Signs operated by electronic or electromechanical controllers located external to the sign shall have a disconnecting means located _____.

(a) within sight of sign   **(b) within sight from controller location**
(c) only in the controller   **(d) only external to the controller**

13. Metallic enclosures of reactors and adjacent metal parts shall be installed so that the _____ from induced circulating currents will not be hazardous to personnel or constitute a fire hazard.

(a) heat   (b) arc   (c) temperature rise   (d) fumes

14. What is the service conductor demand load for the following 20 outlets at a marina?

17 - 30 amp receptacles
 3 - 50 amp receptacles

(a) 660 amp   (b) 528 amp   (c) 462 amp   (d) 396 amp

15. The maximum number of #12 RH conductors that may be installed in a 1 1/2" x 4" octagonal box containing a fixture stud is _____.

(a) 5   (b) 6   (c) 9   (d) 7

16. Battery pack units may supply emergency power if connected _____.

(a) on the lighting circuit of the area   **(b) on any receptacle circuit**
(c) on any branch circuit   **(d) ahead of the main**

17. Which of the following statements about the connection of an outdoor receptacle outlet at a dwelling is (are) correct?

I. the outdoor receptacle outlet may be connected to a general purpose branch circuit
II. the outdoor receptacle outlet may be connected to one of the required small appliance circuits

(a) I only   (b) II only   (c) both I and II   (d) neither I nor II

18. Silicone rubber insulated fixture wire SF-1 should be limited to use where the voltage does not exceed _____ volts.

(a) 500   (b) 300   (c) 200   (d) 100

19. On circuits of 600 volts or less, overhead spans up to 50 feet in length shall have conductors not smaller than _____.

(a) #14   (b) #12   (c) #6   (d) #10

20. A pool recirculating pump motor receptacle shall be permitted not less than _____ feet from the inside walls of the pool.

(a) 5   (b) 8   (c) 10   (d) 15

21. A bare #4 conductor may be concrete-encased and serve as the grounding electrode when at least _____ feet in length.

(a) 25   (b) 15   (c) 10   (d) 20

22. For limited flexibilty for motor connections in a Class I, Division II location, flexible conduit _____.

(a) must be explosion-proof
(b) must be liquid-tight flexible conduit or equal
(c) may be standard flexible metal conduit
(d) shall not be used

23. Flexible metal conduit shall be secured at which of the following?

(a) at intervals not exceeding 4 1/2 feet
(b) within 12 inches on each side of a box where fished
(c) where fished
(d) lengths not exceeding 3' at motor terminals

24. The ampacity of capacitor circuit conductors shall not be less than ____ percent of the rated current of the capacitor.

(a) 100   (b) 125   (c) 135   (d) 150

25. In data processing rooms, the disconnecting means shall disconnect ____.

(a) the A/C, power and lighting to all electronic equipment in the room
(b) the A/C to the room and the power to all electronic equipment in the room
(c) only the power to all electronic equipment in the room
(d) only the equipment operating at below 600 volts

26. When bare grounded conductors are used with insulated conductors, their ampacities are limited to ____.

(a) 60 degrees C
(b) 75 degrees C
(c) 90 degrees C
(d) that permitted for the adjacent insulated conductors

27. Fixtures shall be so constructed or installed that adjacent combustible material will not be subjected to temperature in excess of ____ degrees C.

(a) 75   (b) 90   (c) 185   (d) 140

28. Metal enclosures for grounding electrode conductors shall be ____.

(a) rigid conduit only   (b) not less than 3/4" in diameter
(c) bonded   (d) electrically continuous

29. Storerooms and similar areas adjacent to aircraft hangars but effectively isolated shall be designated ____.

(a) Class I, Division II   (b) Class II, Division I
(c) Class II, Division II   (d) nonhazardous

30. With consideration to mobile homes, which of the following other than a built-in dishwasher are not considered portable appliances if cord connected?

(a) refrigerators   (b) gas cooking appliances
(c) clothes washers   (d) electric ranges

31. Power feed, grounding connection, and shield system connection between the FCC system and other wiring systems shall be accomplished in a ____.

(a) transition assembly   (b) raceway   (c) trench   (d) none of these

32. Audible and visual signal devices shall be employed, where practicable, to give warning of derangement of emergency power.

(a) true   (b) false

33. Type MC cable installed outside of buildings or as aerial cable shall comply with NEC Article ____.

(a) 225   (b) 710-3   (c) 300-5   (d) 225 and 321

34. A ground ring encircling the building, in direct contact with the earth at a depth below earth surface shall not be less than ____ feet, consisting of at least 20' of bare copper not smaller than #2.

(a) 2   (b) 2 1/2   (c) 3   (d) 4

35. Each patient bed in a critical care area shall be provided with a minimum of ____ receptacle(s).

(a) one duplex   (b) single   (c) two duplex   (d) six

36. Service drops not over 300 volts over residential driveways and commercial areas, such as parking lots and drive-in establishments not subject to truck traffic shall have a minimum clearance of ____ feet.

(a) 10   (b) 12   (c) 15   (d) 18

37. Rigid conduit shall be ____ every 10 feet as required by section 110-21.

(a) stamped   **(b) clearly and durably identified**
(c) marked   **(d) none of these**

38. The grounding electrode conductor shall be ____ and shall be installed in one continuous length without a splice or joint.

I. solid   II. solid or stranded   III. insulated, covered or bare

(a) I only   (b) I and III   (c) II and III   (d) III only

39. Branch circuits supplying two or more outlets for fixed space heating equipment in a dwelling shall be rated at ____ amperes.

(a) 15-20-25-30   (b) 15-20-30-40   (c) 15-20-30   (d) 20-30-40

40. Field bends or modifications shall be so made that the ____ of the cable tray system and support for the cables shall be maintained.

(a) temperature   (b) electrical continuity   (c) strength   (d) rigidity

41. Service entrance conductors may be installed before the manufactured building is erected at the site.

(a) true   (b) false

42. The words "thermally protected" appearing on the nameplate of a motor or motor compressor indicate that the motor is provided with a ____.

(a) fuse   (b) breaker   (c) thermal protector   (d) switch

43. Hanging fixtures where located directly above any part of the bathtub shall be so installed that the fixture is not less than ____ feet above the top of the bathtub.

(a) 4   (b) 6   (c) 8   (d) none of these

44. Conductor overload protection is not required if ____.

(a) conductors are oversized by 125%
(b) conductors are part of a limited-energy circuit
(c) interruption of the circuit can create a hazard
(d) none of the above

45. The largest stranded conductor permitted to be connected to terminals by means of upturned lugs is ____ AWG.

(a) #8   (b) #6   (c) #10   (d) #12

46. Electrical installations in hollow spaces, vertical shafts, and ventilation or air-handling ducts shall be so made that the possible spread of fire or products of combustion will not be ____.

(a) substantially increased   (b) allowed   (c) exposed   (d) underrated

47. A branch circuit feeding a sign which has a combination of lamps and transformers shall not exceed the rating of _____ amps.

(a) 15   (b) 20   (c) 30   (d) 50

48. The minimum length of free conductor left at each outlet and switch point in a dwelling shall not be less than _____ inches.

(a) 4   (b) 6   (c) 8   (d) 10

49. Knob and tube wiring splices shall be _____ unless approved devices are used.

(a) taped   (b) bolted   (c) clamped   (d) soldered

50. Each commercial building and each commercial occupancy accessible to pedestrians shall have at least one outside sign outlet branch circuit rated at _____ amps.

(a) 15   (b) 20   (c) either (a) or (b)   (d) neither (a) not (b)

51. Open individual service conductors in wet locations, maximum 600 volts, supported every 9 feet should have a minimum clearance between conductors of _____.

(a) 2 1/2"   (b) 3"   (c) 6"   (d) 12"

52. Angle-pull dimensional requirements apply to junction boxes only when the size of conductor is equal to or larger than _____.

(a) #0   (b) #4   (c) #3/0   (d) #6

53. According to the Code, conductors or poles, where not placed on racks or brackets, shall be supported not less than _____.

(a) 6"   (b) 12"   (c) 18"   (d) 24"

54. Wood braces used in structural mounting of boxes shall have a cross-section not less than nominal _____.

(a) 3/4" x 1 1/2"   (b) 1" x 1 1/2"   (c) 1" x 2"   (d) 3/4" x 2"

55. Pull-type canopy switches shall not be located more than _____ inches from the center of the canopy.

(a) 1 1/2"   (b) 2"   (c) 3"   (d) 3 1/2"

56. Plug fuses of the Edison-base type shall be used ____.

(a) where overfusing is necessary
(b) only as replacement items in existing installations
(c) as a replacement for type S fuses
(d) only for 50 amps and above

57. Where motors are provided with terminal housing, the housing shall be ____ and of substantial construction.

(a) plastic   (b) metal   (c) (a) or (b)   (d) neither (a) nor (b)

58. A storage battery supplying emergency lighting and power shall maintain not less than 87 1/2 percent of full voltage at total load for a period of at least ____.

(a) 2 hours   (b) 1 1/2 hours   (c) 1 hour   (d) 1/4 hour

59. The internal depth of outlet boxes intended to enclose flush devices shall be at least ____.

(a) 1/2"   (b) 7/8"   (c) 15/16"   (d) 1"

60. ____ cable is a factory assembly of two or more insulated conductors in an extruded core of moisture-resistance, flame-retardant, non-metallic material.

(a) NM   (b) NMC   (c) SNM   (d) VSE

61. When supplying nominal 120 volt rated room air-conditioner, the length of the flexible supply cord shall not exceed ____ feet.

(a) 4   (b) 6   (c) 8   (d) 10

62. Auxiliary gutters shall be supported throughout their entire length not exceeding ____ feet.

(a) 5   (b) 6   (c) 8   (d) 10

63. Each general care patient bed location shall be provided with a minimum number of receptacles such as ____.

(a) one single or one duplex
(b) six single or three duplex
(c) two single or one duplex
(d) four single or two duplex

64. Insulating bushings are required on conduit entering boxes, gutters, etc. if it contains conductors as large as _____.

(a) #2   (b) #4   (c) #0   (d) #6

65. Maximum overcurrent protection in amps for a Class 2 remote-control circuit supplied from a power source with overcurrent protection shall be _____ amps.

(a) 1.0   (b) 3.0   (c) 5.0   (d) 7.0

66. A fixture that weighs more than 6 pounds or exceeds _____ inches in any dimension shall not be supported by the screw shell.

(a) 12   (b) 16   (c) 18   (d) 24

67. Bored holes in wood members for cable or raceway-type wiring shall be bored so that the edge of the hole is not less than _____ inches from the nearest edge.

(a) 1 1/4"   (b) 1 1/8"   (c) 1 1/2"   (d) 1 1/16"

68. Unless specified otherwise, live parts of electrical equipment operating at _____ volts or more shall be guarded.

(a) 12   (b) 15   (c) 50   (d) 24

69. Generators operating at _____ volts or less and driven by individual motors shall be considered as protected by the overcurrent device protecting the motor if these devices will operate when the generators are delivering not more than 150% of their full load rated current.

(a) 65   (b) 70   (c) 75   (d) 85

70. Where necessary, a shallow outlet box not less than _____ inches deep may be used.

(a) 1/2   (b) 3/4   (c) 1   (d) 1 1/4

OPEN BOOK
EXAM
#7

70 QUESTIONS
TIME LIMIT - 2 HOURS

TIME SPENT [ ] MINUTES

SCORE [ ] %

1. AC circuits of less than 50 volts shall be grounded under which of the following?

I.   Where installed as overhead conductors outside of buildings.
II. Where supplied by transformers if the transformer supply system is ungrounded.
III. Where supplied by transformers if the transformer supply system exceeds 150 volts to ground.

(a) I only   (b) II only   (c) III only   (d) I, II or III

2. Which of the following conditions must be met for vehicle mounted generators used to supply electrical tools on a construction site?

I. The frame of the generator is bonded to the vehicle frame.
II. The generator supplies cord and plug connected equipment thru receptacles mounted on the
    vehicle or on the generator.

(a). I only   (b) II only   (c) both I and II   (d) neither I nor II

3. Distribution systems for mobile home parks shall be _____.

(a) 120/240v three-phase   (b) 120/208v three-phase
(c) 120/240v single-phase   (d) 120/230v single-phase

4. Metal oxide surge arrester ratings are based on the magnitude and duration of overvoltage at the arrester location as affected by _____.

I. switching surges   II. system grounding techniques   III. phase-to-ground faults

(a) I only   (b) II only   (c) III only   (d) I, II and III

5. _____ devices providing equivalent overcurrent protection in closed-loop power distribution systems shall be permitted as a substitute for fuses or circuit breakers.

(a) Approved   (b) Listed   (c) Accessible   (d) Automatic

6. A unit or assembly of units or sections, and associated fittings, forming a rigid structural system used to support cables is a _____.

(a) flat cable assembly   (b) wireway   (c) multioutlet assembly   (d) cable tray

7. _____ is the distance measured along the enclosure wall from the axis of the centerline of the terminal to a line passing through the center of the opening in the enclosure.

(a) Offset   (b) Radius   (c) Center point   (d) none of these

8. Which of the following is permitted to support a lighting fixture weighing over 50 pounds?

(a) the screwshell of a lampholder   (b) raceway fitting approved for the purpose
(c) an outlet box   (d) fixture wires #14 and larger

9. Which of the following statements about junction boxes is/are true?

I. All shall have a cover   II. All over 6' in length shall have conductors cabled or racked

(a) I only   (b) II only   (c) both I and II   (d) neither I nor II

10. Branch circuits in a recreational vehicle may be derived from autotransformers if _____.

I. the transformer is listed for vehicle use   II. if UL and CSA listed and approved

(a) I only   (b) II only   (c) both I and II   (d) neither I nor II

11. All of the following about paralleling conductors are true except _____.

(a) must terminate in the same manner   (b) must be same material
(c) must be same length   (d) must be enclosed in the same raceway

12. MI cable shall not be bent more than _____.

(a) 2.5" for 1/2" cable   (b) 7" for 1" cable
(c) 6" for 1" cable   (d) 4" for 1/2" cable

13. All of the following may be used on services of 2300/4600v except _____.

(a) MI cable   (b) SE cable   (c) cable bus   (d) busway

14. Nonpower-limited fire protective signaling circuits shall _____.

I.   not be more than 600 volts
II. not exceed 7 amps overcurrent protection for #18 conductor
III. be permitted in same raceway whether AC or DC current

(a) I only   (b) I and II only   (c) II only   (d) I, II and III

15. Wire mesh or other conductive elements where provided in the concrete floor of animal confinement areas to provide an equipotential plane shall be bonded to the building grounding electrode system. The bonding conductor shall be copper, insulated, covered or bare, not smaller than # _____.

(a) 8   (b) 6   (c) 4   (d) 2

16. The provisions of section 300-20 shall not apply to the installation of a _____ conductor in a ferromagnetic envelope (metal enclosure).

(a) grounded   (b) single   (c) insulated   (d) buried

17. A transformer vault ventilation opening shall be located _____ from doors, windows, fire escapes, and combustible material.

(a) 3 feet   (b) 4 feet   (c) 4 1/2 feet   (d) as far away as possible

18. Concealed knob-and-tube wiring shall not be used in the hollow spaces of walls, ceilings and attics where such spaces _____.

(a) exceed 30 degrees C
(b) are insulated by loose or rolled insulation material
(c) are not fire rated for 3 hours
(d) are not ventilated

19. The supply cord conductors and internal wiring of portable high-pressure spray washing machines shall have _____ protection for personnel.

(a) thermal overloads   (b) current limiting fuses
(c) factory installed GFCI   (d) inverse-time breakers

20. The bending radius for nonshielded cables operating at 4160v shall be less than _____ times the diameter.

(a) 6   (b) 8   (c) 10   (d) 12

21. On a delta three-phase, 4-wire system, how many hot wires may use a common neutral?

(a) 2   (b) 3   (c) 4   (d) 6

22. Exposed noncurrent-carrying metal parts of fixed equipment likely to become energized shall be grounded where within _____ feet vertically or 5 feet horizontally of ground or grounded metal objects and subject to contact by persons.

(a) 8   (b) 10   (c) 12   (d) 15

23. Totally enclosed motors of Types (2) or (3) shall have no external surface with an operating temperature in degrees Celsius in excess of _____ percent of the ignition temperature of the gas or vapor involved.

(a) 100% contained   (b) 80   (c) 75   (d) none of these

24. Listed Christmas tree and decorative lighting outfits shall be permitted to be smaller than _____.

(a) #18 AWG   (b) #20 AWG   (c) #22 AWG   (d) #26 AWG

25. All boxes and enclosures for emergency circuits shall be marked so they will be _____ as a component of an emergency circuit.

(a) readily identified   (b) recognized   (c) easily sighted   (d) classified

26. For the purpose of this article, the rating of an adjustable trip circuit breaker having _____ means for adjusting the trip setting shall be the maximum permissible rating or setting.

(a) external means for adjusting   (b) an isolated
(c) readily accessible external   (d) accessible

27. Each section, panel, or strip carrying a number of infrared lampholders shall be considered a(an) _____.

(a) light fixture   (b) appliance   (c) receptacle   (d) outlet

28. This type equipment shall carry a prominent, permanently installed warning regarding the necessity for this grounding feature.

(a) Class I Division I motor equipment
(b) electrostatic equipment
(c) all service entrance equipment in excess of 1200 amps
(d) Class II, Division I service equipment

29. Only wiring methods consisting of _____ shall be installed in ducts or plenums used for environmental air.

I. EMT   II. type NMC   III. type MI   IV. flexible metallic tubing

(a) I and II only   (b) I, II and III only   (c) I, III and IV only   (d) I, II, III and IV

30. If the maximum value of an AC voltage is 170 volts, the R.M.S. value would be aproximately _____ volts.

(a) 294   (b) 98.15   (c) 170   (d) 120

31. For emergency systems, the authority having jurisdiction shall conduct or witness a test on the complete system upon installation and periodically afterward. A _____ shall be kept of such tests.

(a) report   (b) log   (c) written record   (d) chart

32. The service conductors shall be connected to the service disconnecting means by _____ or other approved means.

I. clamps   II. pressure connectors

(a) I only   (b) II only   (c) both I and II   (d) neither I nor II

33. Individual showcases, other than fixed, shall be permitted to be connected by flexible cord to permanently installed receptacles. The installation shall comply with which of the following?

I.   the wiring will not be exposed to mechanical damage
II. attachment plugs shall be grounding type
III. flexible cord shall be hard-service type

(a) I only   (b) II only   (c) III only   (d) I, II and III

34. The grounded conductor of a mineral-insulated, metal-sheathed cable shall be identified at the time of installation by _____ marking at its termination.

(a) distinctive   (b) neutral   (c) solid   (d) identified

35. The prime mover of an emergency generator set _____.

I.   must be provided with an automatic means for starting
II. must be provided with an automatic means of transferring from one fuel supply to another, where dual supplies are used.
III. must have an on-site fuel supply sufficient to operate the prime mover at full demand for 2 hours

(a)) I only   (b) II only   (c) III only   (d) I, II and III

36. Where a service mast is used for the support of service drop conductors, it shall be of adequate strength or be supported by _____.

(a) studs   (b) braces or guys   (c) rigid conduit   (d) R.C. beams

37. For open conductors of not over 600 volts, clearance from ground shall be _____ feet, above finished grade where the supply conductors have a nominal voltage limited to 150 volts to ground and accessible to pedestrians only.

(a) 10   (b) 12   (c) 8   (d) 15

38. The following pool equipment shall be grounded _____.

I. ground-fault circuit-interrupters
II. transformer enclosures
III. electric equipment located within 5 feet of the inside wall of the pool

(a) III only   (b) II and III only   (c) II only   (d) I, II and III

39. The minimum wire bending space within the enclosure for motor controllers for a #1/0 conductor (2 wires per terminal) would be _____ inches.

(a) 6   (b) 5   (c) 4   (d) 2

40. It is the intent of the Code that _____ wiring or the construction of equipment need not be inspected at the time of installation of the equipment, if the equipment has been listed by a qualified electrical testing laboratory.

(a) factory-installed internal   (b) factory-installed
(c) underground   (d) raceway

41. Any motor application shall be considered as _____ unless the nature of the apparatus it drives is such that the motor will not operate continuously with load under any condition of use.

(a) short-time duty   (b) varying duty
(c) continuous duty   (d) periodic duty

42. An overcurrent trip unit of a circuit shall be connected in series with each _____.

(a) ungrounded conductor   (b) grounded conductor
(c) overcurrent device   (d) transformer

43. The statement "Overcurrent protection provided at machine supply terminals" when stamped on a machine tool nameplate means _____.

(a) the overcurrent protection shall be installed by the contractor for the branch circuit to the machine
(b) that thermal protection is built into the machine motors
(c) that provision has been made in the machine tool for each set of supply conductors to terminate in a single circuit breaker or set of fuses
(d) that the contractor must determine the service size based on section 430-62

44. Traveling cables shall be supported by looping the cables around supports for unsupported lengths less than _____ feet.

(a) 300   (b) 200   (c) 150   (d) 100

45. Optional standby systems are _____.

(a) covered under Article 700 emergency systems
(b) not covered in the Code
(c) covered under Article 701
(d) covered in their own section

46. In all cases the work space in front of electrical equipment shall permit at least a _____ degree opening of equipment doors or hinged panels.

(a) 60   (b) 90   (c) 120   (d) 180

47. The N.E.C. covers _____.

I. electric organs   II. speech-input systems   III. centralized sound systems

(a) I only   (b) II only   (c) III only   (d) I, II and III

48. The protective devices(s) shall be capable of detecting and interrupting all values of current which can occur at their location in excess of their trip setting or _____.

I. boiling point   II. melting point   III. capacity

(a) I only   (b) II only   (c) III only   (d) I, II and III

49. A system (of two-wire DC) operating at 50v or less between conductors shall _____.

(a) not be grounded   (b) be grounded
(c) not be permitted   (d) not required to be grounded

50. Mobile x-ray equipment is mounted on a _____ base with wheels and/or casters for moving while completely assembled.

(a) portable   (b) transportable   (c) permanent   (d) temporary

51. In a balanced 3 ø load, the current through any one of the phase loads is always the _____ current.

(a) phase   (b) line   (c) secondary   (d) primary

52. An askarel-insulated transformer installed in a poorly ventilated place shall be furnished with
_____.

I.   a means for absorbing any gases generated by arcing inside the case
II. the pressure-relief vent shall be connected to a flue that will carry such gases outside the building
III. the pressure-relief vent shall be connected to a chimney that will carry such gases outside the
     building

(a) I or II only   (b) I or III only   (c) I only   (d) I, II or III

53. Sign _____ shall be enclosed in metal boxes.

I. flashers   II. fittings   III. cutouts

(a) I and II only   (b) II and III only   (c) I and III only   (d) I, II and III

54. The terminals of an electric-discharge lamp shall be considered energized where any lamp
terminal is connected to a circuit of over _____ volts.

(a) 100   (b) 200   (c) 300   (d) 500

55. Legally required standby systems are typically installed to serve loads such as _____.

I. sewerage disposal   II. data processing   III. refrigeration systems

(a) I and II only   (b) II and III only   (c) I and III only   (d) I, II and III

56. Cases or frames of current transformers, the primaries of which are not over 150 volts to ground
and which are used exclusively to supply current to meters _____.

(a) need to be grounded   (b) need to be isolated
(c) need to be insulated   (d) need _not_ be grounded

57. _____ shall be permitted to be installed in concrete, in direct contact with the earth, or in areas
subject to severe corrosive influences where protected by corrosion protection and judged suitable
for the condition.

(a) PVC   (b) Ceramic   (c) Orangeburg   (d) Rigid metal conduit

58. Connections from headers to cabinets and other enclosures in cellular concrete floor raceways,
shall be made by means of _____ raceways and approved fittings.

(a) rigid nonmetallic   (b) metal   (c) non-metallic   (d) all of these

59. All heating elements that are _____ and are a part of an electric heater shall be legibly marked with the ratings in volts and watts, or in volts and amperes.

I. rated over one amp   II. replaceable in the field   III. a part of an appliance

(a) I only   (b) II only   (c) III only   (d) I, II and III

60. Sign lampholders - 600 volt, nominal or less shall _____.

I. be of the unswitched type having bodies of suitable insulating material
II. be so constructed and installed to prevent turning
III. the screw-shell of all sign lampholders shall be connected to the grounded conductor of the circuit

(a) I, II and III   (b) I and II only   (c) II and III only   (d) I and III only

61. Air conditiong and refrigerating equipment the value of branch-circuit selection current will always be _____ the marked rated-load current.

(a) same as   (b) greater than   (c) less than   (d) same as or greater than

62. A 1 1/2" conduit, 30" long containing over 2 conductors may be filled to _____ square inch.

(a) 1.224   (b) 1.122   (c) 0.816   (d) 0.6324

63. You are to install a 3/4" flexible metallic tubing in an area so that after installation its use will be infrequent flexing, the radius of bends shall not be less than _____ inches.

(a) 3 1/2   (b) 4   (c) 12 1/2   (d) 17 1/2

64. A 76 cubic inch junction box, has two 3/4" rigid conduits entering the bottom of the box, the conduits are supported 36" from the box, select the correct statement.

(a) this box may be supported by wood braces of not less than 3/4" x 1 1/2"
(b) this box must be rigidly supported by a structural member
(c) this box is not required to have any other support
(d) this box may be supported by nails if located within 2" of the back or ends

65. Gas tubing shall be supported independently of the conductors by means of insulators of _____.

I. porcelain   II. glass   III. or by suspension from suitable wires or chains

(a) I, II or III   (b) I or II   (c) II or III   (d) I or III

66. The frames of stationary motors are grounded under which of the following conditions?

I.   where supplied by metal-enclosed wiring
II. in a wet location and not isolated or guarded
III. effectively isolated from the ground
IV. if the motor operates with any terminal at over 150 volts to ground

(a) I and II only   (b) I, II and IV only   (c) III only   (d) I, II, III and IV

67. Where the conductors are fastened to pin-type lampholders that protect the terminals from the entrance of water, the conductors shall be of the _____ type.

I. copper solid   II. aluminum solid   III. copper stranded   IV. aluminum stranded

(a) I only   (b) II only   (c) III only   (d) IV only

68. In a 1500 sq.ft. house, the minimum feeder neutral load for a 5 kva clothes washer/dryer would be _____ kva.

(a) 5.0   (b) 4.3   (c) 3.5   (d) 3.0

69. Formal interpretations of the Code may be found in the _____.

(a) National Electrical Code Handbook
(b) OSHA Standards
(c) NFPA Regulations Governing Committee Projects
(d) Life and Safety Handbook

70. The allowable area to be filled of a 2" conduit is _____ sq.in., if it contains 3 or more conductors.

(a) 2.067   (b) 3.36   (c) 0.8268   (d) 1.34

OPEN BOOK
EXAM
#8

70 QUESTIONS
TIME LIMIT - 2 HOURS

TIME SPENT ☐ **MINUTES**

SCORE ☐ %

1. A 20 ampere receptacle providing shore power for a yacht shall be _____.

(a) GFCI
(b) single and of the locking and grounding type
(c) tamperproof
(d) able to be locked in the "off" position

2. Which of the following pool parts are required to be bonded together?

I.   all fixed metal parts within 5 feet of the inside walls
II. all forming shells
III. all metal parts of an underwater sound system

(a) I only   (b) II only   (c) I and II only   (d) I, II and III

3. When installing office furnishings, receptacle outlets, _____ be located in lighting accessories.

(a) single-type only can   **(b) duplex-type only can**
(c) shall   **(d) shall not**

4. Elevators shall have a single means for disconnecting all ungrounded main power supply conductors for each unit; _____.

(a) this does not include the emergency power service if the system is automatic
(b) this does include the emergency power service
(c) this does not include the emergency power service
(d) no elevators are to operate on emergency power systems

5. A motor operating at 480 volts, the minimum line terminal housing spacing between fixed line terminals must _____ inches.

(a) 1/4   (b) 1/2   (c) 3/8   (d) 5/8

6. What is the minimum thickness for a 6" x 4" x 3 1/4" box?

(a) .0625"   (b) .0747"   (c) 15 MSG   (d) 16 MSG

7. Capacitors containing more than _____ gallons of flammable liquid shall be enclosed in vaults.

(a) 3   (b) 5   (c) 7   (d) 10

8. The branch circuit overcurrent devices in emergency circuits shall be ____.

(a) of the reset type only
(b) a slow-blow type
(c) accessible to only authorized personnel
(d) painted yellow

9. An electrically operated organ shall have both the generator and motor frames grounded or ____.

(a) the generator and motor shall be effectively insulated from ground and from each other
(b) the generator and motor shall be effectively insulated from ground
(c) the generator shall be effectively insulated from ground and from the motor driving it
(d) both the generator and motor shall have double insulation

10. Cable trays include fittings or other suitable means for ____.

I. temperature
II. electric continuity
III. changed in direction and elevation of runs

(a) I only   (b) I and II only   (c) III only   (d) I and III only

11. A raceway contains 45 current-carrying conductors. The ampacity of each conductor shall be reduced ____ percent. (No diversity)

(a) 80   (b) 70   (c) 60   (d) 35

12. Multiconductor portable cables used to connect mobile equipment and machinery above 2000 volts, the conductors shall be ____.

(a) without ground   (b) unshielded   (c) shielded   (d) MK braid

13. A pool panelboard, not part of the service equipment, shall have a grounding conductor installed between ____.

(a) its grounding terminal and a separate ground
(b) its grounding terminal and a ground rod
(c) its grounding terminal and the grounding terminal of the service equipment
(d) its grounding terminal and bonding grid

14. In communication circuits the bonding together of all separate electrodes ____.

(a) shall not be permitted
(b) shall be permitted with a minimum size jumper #4
(c) shall be permitted with a minimum size jumper #6
(d) shall be permitted with a minimum size jumper #8

15. Open motors with commutators shall be located so sparks cannot reach adjacent combustible material, but this _____.

(a) is only required for over 600 volt motors
(b) shall not prohibit these motors on wooden floors
(c) does not prohibit these motors from a Class I location
(d) none of these

16._____ shall be controlled by an externally operable switch or breaker which will open all ungrounded conductors.

I. Outline lighting installation   II. Fixed signs   III. Portable signs

(a) I only   (b) II only   (c) III only   (d) I and II only

17. A public address system _____.

(a) is not covered in the Code　　　　　　**(b) has its own Code section**
(c) is covered in the Code under Art. 640　　**(d) none of these**

18. 3" rigid nonmetallic conduit has a maximum spacing between supports of _____ feet.

(a) 3   (b) 5   (c) 6   (d) 8

19. Conductors which supply one or more AC transformers or DC rectifier arc welders shall be protected by an overcurrent device rated or set at not more than _____ percent of the conductor rating.

(a) 70   (b) 80   (c) 125   (d) 200

20. The battery voltage computed on the basis of _____ volts per cell for the lead-acid type and _____ volts per cell for the alkali type.

(a) 1.5 for lead-acid, 2.0 for the alkali
(b) 2.0 for lead-acid, 1.5 for the alkali
(c) 2.0 for lead-acid, 1.2 for the alkali
(d) 1.2 for lead-acid, 2.0 for the alkali

21. Optional standby systems are typically installed to provide an alternate source of electric power for such loads as _____.

I. communications systems   II. data processing   III. refrigeration

(a) III only   (b) I and III only   (c) II and III only   (d) I, II and III

22. The grounding conductor shall be equal in size to the supply conductors in each cord of an electrically driven irrigation machine, but not smaller than ____copper.

(a) #14   (b) #12   (c) #10   (d) #8

23. Which of the following is **not** a true statement concerning an equipment grounding conductor?

(a) **Under certain conditions, equipment grounding conductors may be required to be larger than circuit conductors.**
(b) **Under certain conditions, equipment grounding conductors may be run in parallel.**
(c) **One size of equipment grounding conductors shall be increased to compensate for voltage drop.**
(d) **One equipment grounding conductor may serve multiple circuits.**

24. In a recreational vehicle, which major appliance, other than built-in does the Code consider portable if cord-connected?

I. refrigerator   II. clothes washers   III. gas range equipment

(a) I only   (b) I and II only   (c) II only   (d) I, II and III

25. A copper bus bar is 4" wide by 1/2" thick. What is the ampacity?

(a) **500 amps**   (b) **1000 amps**   (c) **1500 amps**   (d) **2000 amps**

26. Where rigid PVC conduit is used as a raceway system in bulk storage plant wiring, the raceway shall include ____.

(a) **sunlight resistant listing**   (b) **an equipment grounding conductor**
(c) **a bushing with double locknuts**   (d) **PVC raceway is not permitted**

27. All devices provided with terminals for the attachment of conductors and intended for connection to more than one side of the circuit shall have ____ properly marked for identification.

(a) **conductors**   (b) **terminals**   (c) **sides**   (d) **none of these**

28. There shall be an air space of at least ____ between walls, back, gutter partition, if of metal, or door of any cabinet, or cut out box and nearest exposed current-carrying parts of devices mounted within the cabinet where the voltage exceeds **251** volts.

(a) **1/4"**   (b) **1/2"**   (c) **1"**   (d) **1 1/2"**

29. When two - one ohm resistors are connected in parallel, the total resistance is _____.

(a) 1 ohm   (b) 2 ohm   (c) 1/2 ohm   (d) cannot be calculated

30. A _____ is a receptacle or vessel in which electrochemical reactions are caused by applying electrical energy for the purpose of refining or producing usable materials.

(a) electrolytic cell   (b) heating boiler   (c) radiant heat   (d) duct heater

31. Underground service conductors carried up a pole must be protected from mechanical injury to a height of at least _____ feet.

(a) 10   (b) 8   (c) 12   (d) 15

32. All operating control and signal circuits used in elevators, dumbwaiters and escalators shall be wired with _____ wire or larger.

(a) #14   (b) #16   (c) #18   (d) #20

33. A crane rail when used as a conductor shall be _____.

I.   grounded by fittings used for suspension of the rail
II. grounded by a bonding conductor to a water pipe
III. effectively grounded at the transformer

(a) I only   (b) I and II only   (c) I and III only   (d) II and III only

34. Threaded boxes not over 100 cubic inches, that contain devices, shall be considered to be adequately supported if two or more circuits are threaded into the box wrenchtight and if each conduit is supported within _____ inches of the box.

(a) 36   (b) 30   (c) 24   (d) 18

35. All lights and receptacles in dressing rooms of theaters shall be controlled by wall switches installed in the _____.

(a) dressing rooms   (b) control room   (c) projection room   (d) stage office

36. A number followed by a/an _____ gives the dimensions of THW in sizes #14 to #8 in Table 5, chapter 9 of the Code.

(a) dagger   (b) asterick   (c) obelisk   (d) cross

37. In a three-phase system with a 4-wire, 208/120v, there exists _____.

(a) 3 ø 208 volt, 3 ø 120 volt, and 1 ø 120 volt
(b) 3 ø 208 volt, 1 ø 120 volt, and 1 ø 208 volt
(c) 3 ø 208 volt, 3 ø 120 volt, and 1 ø 208 volt
(d) 1 ø 208 volt, 3 ø 208 volt, and 3 ø 120 volt

38. In general, Class II control circuits and power circuits _____.

I. may occupy the same raceway   II. shall be installed in different raceways

(a) I only   (b) II only   (c) I and II   (d) none of these

39. A boiler employing resistance-type heating elements rated more than 48 amperes and not contained in an ASME rated and stamped vessel shall have the heating elements subdivided into loads not exceeding _____ and protected at not more than _____ amperes.

a 24...48   (b) 48...60   (c) 24...60   (d) 48...80

40. Electrical nonmetallic tubing is permitted to be used in sizes up to _____.

(a) 1"   (b) 2"   (c) 3"   (d) 4"

41. In agricultural buildings all cables shall be secured within _____ inches of a box.

(a) 6   (b) 8   (c) 12   (d) 18

42. Because aluminum is not a magnetic metal, there will be _____ present when aluminum conductors are grouped in a wireway.

(a) no heat due to voltage   (b) no heating due to hysteresis
(c) no induced currents   (d) none of these

43. A residence has a front entrance on the north side of the house along with an attached garage with an 8' wide door, also a back entrance door on the south side of the house. How many lighting outlets are required for these outdoor entrances?

(a) 1   (b) 2   (c) 3   (d) none of these

44. Heating panels or heating panel sets in poured concrete shall not exceed 33 watts per square foot of heated area or _____ watts per linear foot of cable.

(a) 10 1/2   (b) 16 1/2   (c) 25   (d) 33

45. Each switchboard, switchboard section, or panelboard, if used as service equipment, shall be provided with ____.

(a) a main bonding jumper   **(b) a power circuit**
(c) a battery charging panel   **(d) a 4-wire delta connected system**

46. When a diesel engine is used as the prime mover of a generator to supply emergency power, how much of site fuel is required?

(a) one-half hour of fuel supply   **(b) one hour of fuel supply**
(c) two hours of fuel supply   **(d) three hours of fuel supply**

47. The length of a type S cord connecting a trash compactor must not exceed ____.

(a) 18"   **(b) 4'**   **(c) 36"**   **(d) 2'**

48. It is mandatory to identfy the high-leg of a three-phase, 4-wire (120/240v) delta system when ____.

I.   terminating in an equipment panel
II. connected to a three-phase motor
III. the nertral conductor is also present

(a) I only   **(b) II only**   **(c) III only**   **(d) I, II and III**

49. Splices and taps shall not be located within fixture ____.

(a) splice boxes   **(b) arms and stems**   **(c) pancake boxes**   **(d) none of these**

50. In normal practice, the NEC discourages the practice of switching the grounded conductor. However, in one of the following cases, the Code makes an exception and actually requires that this be adopted. Select that case.

(a) electrical circuits that lead or pass through wall mounted fixtures mounted at a height of less than 5' 6" and located in pediatric wards
(b) electrical circuits that lead into or pass through wet niche light fixtures located in swimming pools
(c) electrical circuits that lead into or pass through a gasoline dispensing pump
(d) electrical circuits that lead into or pass through paint spray booths

51. Where practicable, a separation of at least ____ feet shall be maintained between any coaxial cable and lightning conductors.

(a) 2   **(b) 2 1/2**   **(c) 4**   **(d) 6**

52. The neutral of a solidly grounded neutral system shall be permitted to be grounded at more than one point for _____.

I. services
II. direct buried portions of feeders employing a bare copper neutral
III. overhead portion installed outdoors

(a) I only   (b) II only   (c) III only   (d) I, II and III

53. Where extensive metal in or on buildings may become energized and is subject to personal contact _____ will provide additional safety.

(a) adequate bonding and grounding   (b) bonding
(c) suitable ground detectors   (d) none of these

54. Color braid of flexible cords used to identify the use of the grounded conductor shall be finished to show a _____ color, and the braid on the other conductor or conductors finished to show a readily distinguishable solid color or colors.

I. white   II. green   III. natural gray   IV. light blue

(a) I and II only   (b) II and IV only   (c) I and III only   (d) III only

55. Switches, flashers, and similar devices controlling transformers shall be either rated for controlling inductive load(s) or have an ampere rating not less than _____ the ampere rating of the transformer.

(a) 100%   (b) 125%   (c) 200%   (d) 300%

56. Solid dielectric insulated conductors operated above 2000 volts in permanent installations shall have ozone-resistant insulation and shall be _____.

(a) covered   (b) protected   (c) shielded   (d) surface mounted

57. The ampacity requirements of x-ray equipment shall be based on _____ percent of the momentary rating of the equipment.

(a) 40   (b) 50   (c) 70   (d) 80

58. A grounding electrode conductor shall not be required for a system that supplies a _____ circuit and is derived from a transformer rated not more than 1000va.

(a) Class I   (b) Class II   (c) Class III   (d) none of these

59. Motion picture projectors are ____.

(a) covered under theater and similar locations in the Code
(b) not covered in the Code
(c) covered in their own section of the Code
(d) part of the section in the Code on motion picture studios

60. DC conductors used for electroplating shall be protected from overcurrent by ____.

I.   a current sensing device which operates a disconnecting means
II. fuses or circuit breakers
III. other approved means

(a) II only   (b) I and II only   (c) II and III only   (d) I, II and III

61. The paralleling efficiency of ground rods longer than ____ feet is improved by spacing greater than 6 feet.

(a) 8   (b) 10   (c) 15   (d) 20

62. Equipment having an open-circuit voltage exceeding ____ volts shall not be installed in dwelling occupancies.

(a) 1000   (b) 460   (c) 600   (d) 208

63. Two-wire DC circuits used in DC system grounding in an integrated electrical system shall be permitted to be ____.

(a) ungrounded   (b) uninsulated   (c) over 600v   (d) none of these

64. Type USE service entrance cable, identified for underground use in a cabled assembly, may have a ____ concentric.

(a) bare copper   (b) covered metal
(c) bare aluminum   (d) covered

65. Each fixture of each secondary circuit of tubing for electric-discharge lighting system, having an open-circuit voltage of 1500 volts, shall have clearly legible marking reading ____.

I. "Caution 1500 volts"   II. "Caution High Voltage"   III. "Danger High Voltage"

(a) I only   (b) II only   (c) III only   (d) I, II and III

66. A _____ is a protective device for limiting surge voltages by discharging or bypassing surge current, and it also prevents continued flow of follow current while remaining capable of repeating these functions.

(a) surge arrester   (b) auto fuse   (c) fuse   (d) circuit breaker

67. Motor branch-circuit, short-circuit and ground fault protection and motor overload protection shall be permitted to be combined in a single protection device where the rating or setting of the device provides the _____ protection specified in section 430-32.

(a) combined overload   (b) overcurrent   (c) overload   (d) branch-circuit

68. The maximum ampacity for a single #500 MCM conductor in a type IGS-EC cable shall not exceed _____ amperes.

(a) 119   (b) 168   (c) 206   (d) 238

69. For cord and attachment, plug-cord connected motor-compressor and equipment on 15 or 20 ampere branch-circuits, the rating of the attachment plug and receptacle shall **not** exceed 20 amperes at _____ volts nor 15 amperes at _____ volts.

(a) 250...125   (b) 125...250   (c) 250...250   (d) 125...125

70. Loop wiring for underfloor raceways, shall not be considered _____.

I. a splice   II. a tap

(a) I only   (b) II only   (c) both I and II   (d) neither I or II

OPEN BOOK
EXAM
#9

70 QUESTIONS
TIME LIMIT - 2 HOURS

TIME SPENT ☐ MINUTES

SCORE ☐ %

1. The minimum size conductor permitted in parallel for elevator lighting is _____, provided the ampacity is equivalent to a #14 wire.

(a) #14   (b) #20   (c) #16   (d) #1/0

2. Conductors of light and power systems of all voltages may occupy the same enclosure or raceway _____.

(a) if less than 600 volts and if insulated for maximum voltage of any conductor within the enclosure or raceway
(b) if power system is over 600 volts and light system under 600 volts
(c) if power system is over 600 volts and individual circuits are AC
(d) in most instances without qualification

3. Transformers rated over _____ KV shall be installed in a vault

(a) 10   (b) 12 1/2   (c) 25   (d) 35

4. Where used outside, aluminum or copper-clad aluminum grounding conductors shall not be installed within _____ inches of earth.

(a) 24   (b) 18   (c) 12   (d) 6

5. Tubing having cut threads and used as arms or stems on light fixtures may not be less than _____ inches wall thickness.

(a) .020   (b) .025   (c) .040   (d) .015

6. The supply leads to sign boxes, cabinets, and outline troughs shall _____.

(a) not be required to be enclosed
(b) be enclosed in metal
(c) be enclosed in noncombustible material
(d) be enclosed in a lead sheath

7. Types TP, TPT, TS and TST shall be permitted in lengths not exceeding _____ feet when attached directly, or by means of a special type of plug, to a portable appliance rated at 50 watts or less.

(a) 8   (b) 10   (c) 15   (d) can't be used at all

8. When determining the load on the "volt-amps per square foot" basis, the floor area shall be computed from the _____ dimensions of the building.

(a) inside (b) outside (c) either (a) or (b) (d) neither (a) nor (b)

9. Parts of electric equipment which in ordinary operation produce _____ shall be enclosed or separated and isolated from all combustible material.

I. molten metal II. flames III. sparks

(a) II only (b) III only (c) I and II only (d) I, II and III

10. Locations where combustible dust is normally in heavy concentrations are designated as _____.

(a) Class I, Division II   (b) Class II, Division I
(c) Class II, Division II   (d) Class III, Division I

11. Which of the following is not a standard classification for a branch circuit supplying several loads?

(a) 20 amp (b) 25 amp (c) 30 amp (d) 50 amp

12. The minimum size conductor for lighting elevator circuits traveling cables is _____.

(a) #12 (b) #18 (c) #16 (d) #14

13. Steel cable trays shall not be used as equipment grounding conductors for circuits protected above _____ amperes.

(a) 200 (b) 60 (c) 600 (d) 1200

14. Conductive materials enclosing electrical conductors are grounded to _____.

II prevent lightning surges
II. prevent voltage surges
III. to aid the overcurrent device for shorts and ground faults

(a) I only (b) II only (c) III only (d) all of these

15. No grounded conductor shall be attached to any terminal or lead so as to reverse designated _____.

(a) phase (b) angle (c) polarity (d) line

16. Input leads to motor-generators used in conjunction with sound-recording equipment shall be ____.

(a) run with the output leads
(b) run separately from the output leads
(c) run in lead sheaths
(d) twisted together to lower impedance

17. When the number of receptacles for an office building is unknown, an additional load of ____ volt-amp(s) per square foot is required.

(a) 1   (b) 2   (c) 1/2   (d) 3

18. If a protective device rating is marked on an appliance, the branch circuit overcurrent device rating shall not exceed ____ the protective device rating marked on the appliance.

(a) at all   (b) more than 50%   (c) 80%   (d) 125%

19. Unless identified for use in the operating environment, no conductors or equipment shall be located in ____ having a deteriorating effect on the conductors or equipment.

I. damp or wet locations
II. where exposed to gases, fumes, vapors, liquids, or other agents

(a) I only   (b) II only   (c) I and II   (d) none of these

20. All theater fixed stage switchboards that are not completely enclosed dead front and dead rear or recessed into a wall, shall be provided with a metal hood extending the full length of the board to protect all equipment from falling objects.

(a) true   (b) false

21. The conduit or raceways, including their end fittings, shall not rise more than ____ inches above the bottom of the enclosure.

(a) 3   (b) 4   (c) 5   (d) 6

22. Service entrance cables, where subject to physical damage, shall be protected in ____.

I. EMT   II. IMC   III. rigid metal conduit

(a) III only   (b) II and III   (c) I, II and III   (d) I and III

23. All buildings or portions of buildings or structures designed or intended as a place of assembly shall have _____ or more persons.

(a) 50   (b) 100   (c) 250   (d) 500

24. _____ and larger grounding electrode conductors shall be protected where exposed to severe physical damage.

(a) #8   (b) #4   (c) #2   (d) #6

25. Where storage batteries are used for emergency systems they shall be _____.

(a) provided with automatic battery charging means
(b) alkali or lead type
(c) neither (a) nor (b)
(d) both (a) and (b)

26. The following letter suffixes shall indicate the following:

_____ for two insulated conductors laid parallel within an outer nonmetallic covering.

(a) D   (b) M   (c) R   (d) N

27. Lead wires on weatherproof lampholders shall not be less than _____ wire.

(a) #12   (b) #14   (c) #16   (d) #10

28. When connections are made in the white wire in a multi-wire circuit at receptacles, they are required to be made _____.

(a) connected to the silver terminals on the duplex
(b) to the brass colored terminal
(c) with a pigtail to the silver terminal
(d) none of these

29. In a grounded system, the conductor that connects the circuit grounded conductor at the service and/or the equipment grounding conductor to the grounding electrode is called the _____.

(a) main grounding conductor
(b) common main grounding conductor
(c) equipment grounding conductor
(d) grounding electrode conductor

30. Remote control conductors for motor controls shall be protected at not more than _____ percent.

(a) 100 (b) 200 (c) 300 (d) 500

31. The overall covering for type NMC cable shall be _____.

I.   flame retardant
II. moisture resistant
III. fungus resistant
IV. corrosion resistant
V.   all of these

(a) II and III (b) I and III (c) II, III and IV (d) V

32. A flush recessed fixture equipped with a solid lens when installed in a closet must:

(a) be on the wall above the closet door at least 18" upward and from a storage area where combustible materials may be stored
(b) be on the ceiling over an area unobstructed to the floor, at least 18" upward and horizontally from a storage area where conbustible materials may be stored
(c) (a) and (b)
(d) none of these

33. Health care facilities, general care patient, how many MV's are exposed raceways allowed to have under normal operation?

(a) 100 (b) 250 (c) 500 (d) 1000

34. The neutral feeder conductor must be capable of carrying the maximum _____ load.

(a) connected (b) unbalanced (c) demand (d) grounded

35. For straight pulls, the length of the box shall be not less than _____ the outside diameter, over sheath, of the largest conductor or cable entering the box on systems over 600 volts.

(a) 8 times (b) 6 times (c) 36 times (d) 48 times

36. Expansion joints and telescoping sections of raceway shall be made electrically continuous by equipment _____ or other means approved for the purpose.

(a) grounding conductors   **(b) grounded conductor**
(c) bonding jumper   **(d) none of these**

37. Sealing compound is employed with mineral-insulated cable in a Class I location for the purpose of _____.

(a) preventing passage of gas or vapor
(b) excluding moisture
(c) limiting a possible explosion
(d) preventing escape of powder

38. A lighting and appliance branch circuit panelboard contains six - 3 pole circuit breakers and eight - 2 pole circuit breakers. The maximum allowable number of single pole circuit breakers permitted to be added is _____.

(a) 8   (b) 16   (c) 28   (d) 12

39. The Code provides that unshielded lead-in conductors of amateur transmitting stations shall clear the building surface which is wired over by a distance not less than _____ inches.

(a) 1   (b) 2   (c) 3   (d) 4

40. A 24 volt landscape lighting system installed with UF cable is permitted with a minimum cover of _____ inches.

(a) 6   (b) 12   (c) 18   (d) 24

41. Each circuit leading to or through a dispensing pump shall be provided with a switch or other acceptable means to disconnect _____ from the source of supply all conductors of the circuit, including the grounded neutral, if any.

(a) automatically (b) simultaneously   (c) manually   (d) individually

42. Floor boxes shall be considered to meet the requirement of the spacing receptacles on walls if they are _____.

(a) within 12" of the wall   (b) within 18" of the wall
(c) close to the wall   (d) none of these

43. For general wiring in Class I, Division I locations it is permissible to use _____.

(a) rigid metal conduit   (b) EMT   (c) flexible metal conduit   (d) all of these

44. A dwelling general lighting load of 4500va requires how many 15 amp circuits?

(a) 2   (b) 3   (c) 4   (d) 5

45. In a commercial garage the pit shall be classified _____ unless provisions are made for six air changes per hour.

(a) Class I, Divison II
(b) Class II, Division II
(c) Class II, Divison I
(d) Class I, Division I

46. Ground clamps shall be approved for general use without protection or shall be protected _____.

I.   by enclosing in wood
II. by enclosing in metal
III. by equivalent protective covering

(a) II only   (b) II and III only   (c) I and III only   (d) I, II and III

47. Each doorway leading into a vault from the building interior shall be provided with a tight fitting door having a minimum fire rating of _____ hours.

(a) 2   (b) 4   (c) 5   (d) 3

48. Auxiliary equipment for electric-discharge lamps shall be enclosed in noncombustible cases and _____.

(a) not over 3" away   (b) not over 1500w
(c) treated as a source of heat   (d) none of these

49. Tap conductors for household cooking equipment supplied from a 50 amp branch circuit shall have an ampacity of not less than _____.

(a) 50   (b) 70   (c) 20   (d) 80

50. The ampacity for the supply conductors for a resistance welder with a duty cycle of 15% and a primary current of 21 amps is _____ amps.

(a) 9.45   (b) 8.19   (c) 6.72   (d) 5.67

51. Insulated wires shall be marked or tagged with which of the following?

(a) maximum working voltage
(b) type letters
(c) manufacturer indentification
(d) all of these

52. The maximum fill in wireways or gutters used for sound-recording installations is _____.

(a) 20%   (b) 58%   (c) 50%   (d) 75%

53. Where a change occurs in the size of the ungrounded conductor _____.

(a) the neutral can be reduced two sizes
(b) the grounded conductor can be reduced 70%
(c) a similar change may be made in the size of the grounded conductor
(d) the only reduction for the neutral is household ranges

54. Which of the following applies to Class I Division I locations?

(a) ignitible flammable gases and vapors
(b) grain silos
(c) ignitible fibers or flyings
(d0 combustible dust

55. Tap devices used in FC assemblies shall be rated at not less than _____ amps or more than 300 volts, and they shall be color-coded in accordance with the requirements of 363-20.

(a) 20   (b) 15   (c) 30   (d) 40

56. The screw shell contact of all sign lampholders in grounded circuits shall be connected to _____.

(a) the white or gray conductor
(b) the grounded conductor
(c) the neutral
(d) any of these

57. Underground service conductors must have a rating not smaller than _____.

(a) #3   (b) #4   (c) #6   (d) #8

58. The Code provides that in Class I, Division I locations conduit seals shall be placed not farther from spark-producing devices than _____ inches.

(a) 18   (b) 24   (c) 12   (d) 30

59. Where conduit is threaded in the field, it is assumed that a standard cutting die should provide _____ taper per foot.

(a) 1/4"   (b) 1/2"   (c) 3/4"   (d) 1'

60. How would you connect the grounded system conductor to the grounding electrode?

(a) grounded conductor　　　　**(b) grounding conductor**
(c) bonding jumper　　　　**(d) bonding jumper main**

61. The height of a circuit breaker used as a switch shall not exceed _____ feet above the floor.

(a) 4　(b) 4 1/2　(c) 5　(d) 6 1/2

62. Equipment intended to break at other than fault levels shall have an interrupting rating system voltage sufficient for the current that must be interrupted.

(a) true　(b) false

63. Ground-fault protection that functions to open the service disconnecting means _____ protect(s) service conductors or the service disconnecting means.

(a) will　(b) will not　(c) adequately　(d) totally

64. Enclosures for isolating switches in Class I, Division II locations _____.

(a) may be of general-use type
(b) must be explosion-proof type
(c) must have interlocking devices
(d) may not have doors in hazardous area

65. Which of the following statements about the protection of NM sheathed cable from physical damage is (are) correct?

I.　when passing through a floor the cable shall be enclosed in pipe or conduit extending at least 6 inches above the floor
II.　when run across the top of the floor joists in an accessible attic, the cable shall be protected by guard strips

(a) I only　(b) II only　(c) both I and II　(d) neither I nor II

66. The type of grounding receptacle or plug required in the hazardous location of an anesthetizing location shall be listed as _____.

(a) special grade receptacle
(b) Class I, Division I type receptacle
(c) Class II, Division II receptacle
(d) Class I, Group C type receptacle

67. Lengths of not more than _____ of AC cable at terminals where flexibility is necessary does not have to follow the surface of the building.

(a) 28"   (b) 2'   (c) 30"   (d) 3'

68. Any room or location in which flammable anesthetics are stored shall be considered to be a _____ area from floor to ceiling.

(a) Class I, Division I   (b) Class I, Division II
(c) Class II, Division II   (d) nonhazardous

69. Circuits containing electric-discharge lighting transformers exclusively shall not be rated in excess of _____ amps.

(a) 20   (b) 30   (c) 40   (d) 50

70. Where raceway-type service masts are used, all raceway fittings shall be _____ for use with service masts.

(a) identified   (b) approved   (c) heavy-duty   (d) none of these

OPEN BOOK
EXAM
#10

70 QUESTIONS
TIME LIMIT - 2 HOURS

TIME SPENT ☐ MINUTES

SCORE ☐ %

1. Exposed live parts within porcelain fixtures shall be suitably recessed and so located as to make it improbable that wires will come in contact with them. There shall be a spacing of at least _____ between live parts and the mounting plane of the fixture.

(a) 1/4"   (b) 1/8"   (c) 1/2"   (d) 3/4"

2. A blue lead in a heating cable means that the voltage is _____ volts.

(a) 120   (b) 208   (c) 240   (d) 277

3. Outlets for specific appliances such as laundry equipment, shall be within _____ feet of the appliance.

(a) 4   (b) 6   (c) 8   (d) 10

4. An enclosure designed either for surface or flush mounting and provided with a frame, mat, or trim in which a swinging door or doors are or may be hung is a _____.

(a) panelboard   (b) switchboard   (c) wireway   (d) cabinet

5. Grounding of a metal raceway used to protect Romex is required if the raceway is _____ feet or over within reach of ground or grounded metal.

(a) 6   (b) 8   (c) 10   (d) 25

6. To prevent the entrance of moisture, service-entrance conductors shall be connected to the service-drop conductors _____.

I. below the level of the termination of the service-entrance cable sheath
II. below the level of the service head

(a) I only   (b) II only   (c) both I and II   (d) neither I nor II

7. The area of square inches for a #6 bare conductor is _____.

(a) 0.017   (b) 0.027   (c) 0.207   (d) 0.184

8. Which of the following is **not** a standard size fuse?

(a) 110 amp   (b) 125 amp   (c) 75 amp   (d) 250 amp

9. In exposed type of show window signs, terminals shall be _____.

(a) suitable guarded by grounded metal enclosure
(b) enclosed by receptacles
(c) inaccessible
(d) supported so as to maintain separation of not less than 2" between wires

10. The N.E.C. covers _____.

I. gas welders II. DC rectifier arc welders III. motor-generator arc welders IV. resistance welders

(a) I and IV only   (b) I, II and III only   (c) II, III and IV only   (d) I, II, III and IV

11. Type FCC cable wiring system is designed for installation under _____.

(a) tile   (b) carpet   (c) carpet squares   (d) concrete

12. Service cables mounted in contact with a building shall be supported at intervals not exceeding _____ feet.

(a) 10   (b) 6   (c) 2 1/2   (d) 4 1/2

13. Multispeed motors shall be marked with the code letter designating the locked-rotor kva horsepower for the highest speed at which the motor _____.

(a) can be stalled　　　　**(b) can be started**
(c) needs to be rated　　　**(d) can run safely**

14. Soft-drawn or medium-drawn copper, lead in conductors for receiving antenna systems shall be permitted where the maximum span between points of support is less than _____ feet.

(a) 35   (b) 30   (c) 20   (d) 10

15. Non-heating leads of heating cables operating in 240 volt systems, shall have a _____ color.

(a) red   (b) blue   (c) yellow   (d) brown

16. A pool with a maximum dimension of _____ feet and a maximum wall height of 42", and is so constructed that it may be readily disassembled for storage and reassembled to its original integrity is a storable pool.

(a) 18   (b) 5   (c) 10   (d) 3

17. Branch circuit conductors within 3" of a ballast, within the ballast compartment shall be recognized for use at temperatures not lower than 90 degrees C, such as insulation types _____.

I. THHN   II. THW   III. TW   IV. FEP

(a) I only   (b) I and IV only   (c) I, II and IV   (d) I, II, II and IV

18. Sign conductors in raceways, metal-clad cable, or enclosures _____ shall be of the lead-cover type or other type specially listed for the conditions.

(a) buried   (b) exposed to weather   (c) in free air   (d) none of these

19. Flexible cord shall be considered as protected by a 20 amp branch circuit overcurrent device if the cord is _____.

(a) not less than 6' in length   **(b) #20 or larger**
(c) #18 or larger   **(d) #16 or larger**

20. Infrared lamps for industrial heating appliances shall have overcurrent protection not exceeding _____ amps.

(a) 30   (b) 40   (c) 50   (d) 60

21. The temperature limitation of MI cable is based on the _____.

(a) ambient temperature   **(b) conductor insulation**
(c) insulating materials used in the end seal   **(d) none of these**

22. Overcurrent devices shall not be located in the vicinity of easily ignitible material such as in _____.

(a) bedrooms   (b) clothes closets   (c) kitchens   (d) garages

23. A new building will have two service heads serviced by one service drop. What is the maximum distance apart that the Code permits these heads to be located?

(a) 36"
(b) 48"
(c) 6 feet
(d) there is no maximum distance as long as the service entrance conductors are long enough to reach the service drop conductors

24. Liquidtight flexible metal conduit is shipped in what sizes minimum and maximum?

(a) 1/2" to 4"   (b) 1/2" to 6"   (c)3/4" to 5"   (d) 1/2" to 2"

25. The rated primary current of an electric resistance welder is _____.

I.   2.8 times the primary current divided by the duty cycle
II. determined by the percentage of time during which the welder is loaded
III. the rated kva multiplied by 1000 and divided by the rated primary voltage, using values given on the nameplate
IV. the current drawn from the supply circuit during each welder operation at the particular heat tap and control setting used

(a) I and III only   (b) II and IV only   (c) I, II and III   (d) I, II, III and IV

26. Renewable type contacts shall be provided on all knife switches rated 600 volts designed for use in breaking current over _____ amperes.

(a) 50   (b) 100   (c) 150   (d) 200

27. Voltage shall not exceed 600 volts between conductors on branch circuits supplying only ballasts for electric-discharge lamps in tunnels with a height of not less than _____ feet.

(a) 12   (b) 15   (c) 18   (d) 22

28. 15 and 20 ampere receptacles located in pediatric areas shall be _____.

(a) tamperproof   (b) isolated   (c) GFI   (d) specification grade

29. A nipple contains six #8 THW copper current-carrying conductors. The ampacity of each conductor would be _____amperes.

(a) 24   (b) 50   (c) 35   (d) 40

30. Type MTW insulation would be used for _____.

(a) switchboards only   (b) machine tool wiring
(c) feeders only   (d) fixtures

31. Which of the following would **not** be approved in all Class II locations?

(a) flexible connections   (b) threaded bosses
(c) dust-tight boxes   (d) EMT

32. Emergency lighting, emergency power or both, in a building or group of buildings will be available within the time required for the application, but not to exceed one of the following:

(a) 5 seconds   (b) 10 seconds   (c) 30 seconds   (d) 60 seconds

33. Aluminum grounding conductors shall not be used where _____.

I. within 18" of earth
II. subject to corrosive conditions
III. in direct contact with masonry

(a) I only   (b) II only   (c) III only   (d) I, II and III

34. The heating of conductors depends on _____ values which, with generator field control, are reflected by the nameplate current rating of the motor-generator set driving motor rather than by the rating of the elevator motor, which represents actual but short-time and intermittent F.L.C. values.

(a) root-mean-square current   (b) voltage
(c) ambient temperature   (d) ozone

35. Each vented cell shall be equipped with a _____ designed to prevent destruction of the cell.

(a) gas arrestor   (b) electrolyte   (c) flame arrestor   (d) hydrometer

36. General power and lighting in a health care facility _____.

(a) are not part of the emergency system
(b) are connected to the alternate source generator
(c) are part of the continuous power system
(d) are part of the equipment system

37. Parking garages used for parking or storage and where no repair work is done, open flame, welding, or the use of volatile flammable liquids are _____.

(a) Class I   (b) Class II   (c) Class III   (d) not classified

38. The maximum weight of a light fixture that may be mounted on the screw shell of a brass socket is _____ pounds.

(a) 2   (b) 6   (c) 8   (d) 10

39. The basic photovoltaic device which generates electricity when exposed to light is a _____.

(a) battery   (b) diode   (c) solar cell   (d) mandrel

40. The National Electrical Code scope includes _____.

(a) conductors and equipment on certain public or private premises
(b) conductors and equipment on all buildings or premises where electrical power is
**   employed**
(c) conductors and equipment on all buildings, premises and conveyances where electrical
**   power is employed**
(d) conductors and equipment wherever and whenever electricity is employed

41. Open core and coil transformers used in electric signs shall be limited to _____ volts.

(a) 2500 (b) 5000 (c) 10000 (d) 12000

42. Fixtures in clothes closets should be installed _____.

(a) 18" from ceiling   **(b) 24" from ceiling**
(c) on ceiling only   **(d) on ceiling or on wall above door**

43. Open conductors on insulators must be covered when they are within _____ feet of a building.

(a) 10 (b) 12 (c) 15 (d) 25

44. Flexible cords shall be connected to devices and to fittings that tension will not be transmitted
to joints or terminal screws. This shall be accomplished by ___.

(a) a knot in the cord   **(b) winding with type**
(c) a special fitting   **(d) all of these**

45. Open conductors run individually as service drops shall be _____.

I. insulated II. bare III. covered

(a) I only (b) II only (c) III only (d) either I or III

46. In patient care areas of a hospital, the equipment grounding terminal bars of the normal and
essential electrical system panelboards shall be bonded together with an insulated continuous copper
conductor not smaller than _____.

(a) #8 (b) #6 (c) #10 (d) #12

47. The radius of a 3" conduit containing conductors with lead sheath shall not be less than _____
inches.

(a) 18 (b) 31 (c) 21 (d) 36

48. All but which of the following are approved wiring methods for hazaradous Class II locations?

(a) threaded boxes   **(b) seal boxes with telescoping covers**
(c) flexible metal conduit   **(d) EMT**

49. FCC cable connections shall use connectors identified for their use, such that _____ against dampness and liquid spillage are provided.

I. sealing   II. insulation   III. electrical continuity

(a) I only   (b) II only   (c) III only   (d) I, II and III

50.The circuit supplying an autotransformer-type dimmer installed in theaters and similar places shall not exceed how many volts between conductors?

(a) 480 volts   (b) 277 volts   (c) 240 volts   (d) 150 volts

51. Steel raceways installed in tunnels shall not be used as an equipment grounding conductor over _____ volts.

(a) 150   (b) 300   (c) 600   (d) 1000

52. Raceways embedded in a masonry wall or buried beneath a floor shall be considered _____.

(a) to be outside the hazardous area if any extensions pass thru such areas
(b) to be within the hazardous area if any connections lead into such areas
(c) to be hazardous if beneath a hazardous area
(d) are never hazardous

53. Surface metal raceways when extended through walls or floors must be in _____ lengths.

(a) 8 foot   (b) 6 foot   (c) 4 foot   (d) none of these

54. For a welder having a time rating of one hour, the rated ampacity of the supply conductor shall be determined by multiplying the rated primary current in amperes by _____ if it is an AC transformer.

(a) .63   (b) .75   (c) .78   (d) .84

55. Compliance with the provisions of the Code will result in _____.

(a) good electrical service   **(b) an efficient system**
(c) freedom from hazard   **(d) all of these**

56. In general, busways shall be supported at intervals of _____ feet.

(a) 3   (b) 5   (c) 6   (d) 10

57. All bonded parts of a pool shall be connected to a common bonding grid with a _____ conductor.

I. #8 stranded   II. #6 solid aluminum   III. #8 solid copper

(a) I only   (b) II only   (c) III only   (d) I or III

58. The Code requires all conductors that attach to a cablebus to be in the same raceway _____.

(a) because its cheaper   (b) because its easier to test
(c) because its easier to maintain   (d) because of inductive current

59. Flexible cords to portable electrically heated appliances rated at more than _____ watts shall be approved for heating cords.

(a) 50   (b) 100   (c) 300   (d) 500

60. Electrical equipment provided with ventilating openings shall be installed so that walls or other obstructions do not prevent the _____ of air through the equipment.

(a) movement   (b) release   (c) free circulation   (d) none of these

61. A motor control circuit _____.

I.   carries electric signals to the controller, and carries the main power
II. does not carry electric signals to the controller, but carries the main power
III. carries the electric signals to the controller, but does not carry main power

(a) I only   (b) II only   (c) III only   (d) none of these

62. What is the maximum length of an unprotected feeder tap conductor?

(a) 10 feet   (b) 25 feet   (c) 50 feet   (d) 100 feet

63. Electric ranges and clothes dryer used in mobile homes shall be _____.

(a) installed with the equipment enclosure ungrounded
(b) installed in the same manner as a dwelling unit
(c) installed with a grounding conductor run in parallel with the neutral
(d) installed with the grounded circuit conductor insulated from the grounding conductor and
    equipment enclosures

64. What is the volt-amp input of a fully loaded 10 hp 208v, three-phase motor?

(a) 5824va   (b) 6406va   (c) 10,087va   (d) 11,096va

65. Non-insulated busbars will have a minumum space of _____ inches between the bottom of enclosure and busbar.

(a) 6   (b) 8   (c) 10   (d)12

66. Wet-niche lighting fixtures shall be connected to an equipment grounding conductor not smaller than _____.

(a) #10   (b) #6   (c) #8   (d) #12

67. The minimum insulation level for neutral conductors of solidly grounded systems shall be _____ volts.

(a) 600   (b) 1000   (c) 1500   (d) 2100

68. The maximum operating temperatures of rubber-covered, type RFH-1 heat resistant fixture wire is _____ degrees F.

(a) 124   (b) 167   (c) 195   (d) 224

69. The #4 solid copper conductor connecting the grounding electrode to the service has been accidently cut. Which of the following could be acceptable connectors for splicing this conductor?

I. a split bolt connector
II. a compression sleeve type connector
III. a wire nut

(a) I only   (b) I or II only   (c) I, II and III   (d) none of these are acceptable

70. Instrument pilot lights and potential current transformers shall be protected by O.C.P. of _____ amp or less.

(a) 15   (b) 20   (c) 30   (d) 50

ANSWERS

1. **(b) 110-19.** Key word is **not** listed in the index. Think of a word similar to "trolley" ...."Railway". Check the index for "Railway" and it will lead you to the answer in 110-19.

2. **(c) 210-6d1(b).** "Ballasts" and "Electric-discharge lamps" are listed in the index but are no help. "Tunnel" is not even listed in the index. The key words "Lighting fixture voltages" from the index will lead you to 210-6 and the answer.

3. **(a)**   This question is not from the NEC, the answer is found in the American Electrician's Handbook. The key is to take other electrical reference books to the exam with you.

4. **(d) 520-42.** "Proscenium" is not listed in the index, "light fixtures" is listed in the index, but no help. "Conductors" and "Insulation" are listed in the index, but it would take a **considerable** amount of **time** to look up each article under "Conductors" and "Insulation". The key is if you know the meaning of the word "proscenium". The dictionary states: Theater stage in front of curtain. The NEC index for "Stage" or "Theater" lists article 520 Part C, stage equipment 520-42. For this type of question it sometimes helps to have a dictionary.

5. **(d) 665-24.** Words "Auxiliary rectifiers", "Bias supplies", "Tube keyers", and "Bleeder resistors" are not listed in the index. Key word is "Capacitor", listed in the NEC index. Now its a matter of checking each listing until you check Induction heating 665-24 which will lead you to the answer.

6. **(d) 640-9b.** The key word is "Batteries". The NEC index states to look under "Storage batteries". There are seven different article listings for storage batteries, the **last** listing "sound recording equipment" is the one that will lead you to the answer in 640-9b.

7. **(a) 640-6b.** "Input leads" not listed in the index. "Motor-generator" listed, but only for arc welders. "Rotary converters" not listed in the index. Here is a simple true or false question that is very difficult to locate in the NEC. The key is reading and retaining; As you read the Code looking for an answer, try to retain as much from each section as you can. By carefully reading article "640 Sound Equipment" to answer the previous question (#6), you would have read the answer to this question in 640-6b.

MASTER CLOSED BOOK EXAM #1

1. **(d)** 120°
2. **(b)** passing a point per second
3. **(d)** hydrometer
4. **(c)** voltmeter
5. **(d)** expansion bolts
6. **(b)** less voltage drop
7. **(a)** remain constant
8. **(c)** average
9. **(b)** voltmeter, ohmmeter, ammeter
10. **(a)** burn at full brightness
11. **(c)** will not   230-95b FPN4
12. **(d)** grounding cond. DEF 100
13. **(a)** iron
14. **(a)** ruin the tip
15. **(a)** hanging fixture
16. **(a)** weatherproof DEF 100
17. **(c)** reverse any two
18. **(d)** cannot be used in e.p.
19. **(c)** three
20. **(c)** capacitor
21. **(d)** 180va
22. **(d)** strip sandpaper
23. **(c)** induction
24. **(b)** current
25. **(a)** Kirchhoff's law
26. **(c)** E = I x R
27. **(d)** clamped perpendicular
28. **(a)** prevent loosening under vibration
29. **(a)** primary
30. **(d)** heat sensing element
31. **(b)** ammeter
32. **(b)** in a DC circuit
33. **(d)** conductor
34. **(a)** shock or injury
35. **(a)** oil
36. **(d)** conductance
37. **(b)** concealed
38. **(b)** autotransformer
39. **(a)** may transmit shock
40. **(b)** heater coils
41. **(c)** the feeder
42. **(a)** general purpose DEF 100
43. **(c)** ease of voltage variation
44. **(a)** parallel
45. **(a)** 0.002v
46. **(b)** horsepower
47. **(d)** stranded 410-28e
48. **(a)** use a template
49. **(c)** of inductance
50. **(b)** 3ø 208v, 1ø 120v and 1ø 208v

1. **(c)** electro
2. **(a)** cutting off
3. **(c)** tachometer
4. **(b)** b.c. protection
5. **(b)** askarel DEF 100
6. **(d)** % of applied voltage
7. **(c)** demand factor DEF 100
8. **(b)** pressure
9. **(d)** either I or II
10. **(c)** less than any one
11. **(c)** wires do not support the light
12. **(d)** I, II & III
13. **(c)** RHH
14. **(c)** simultaneously
15. **(c)** area surrounding
16. **(d)** both a & b
17. **(b)** pf meter
18. **(c)** covered DEF 100
19. **(d)** I,II,III & IV
20. **(d)** all DEF 100
21. **(d)** varying
22. **(b)** low transformation
23. **(d)** Δ
24. **(d)** thermal cutout
25. **(b)** short circuited
26. **(c)** ohms
27. **(c)** volt
28. **(a)** current flow develops heat
29. **(c)** 90-2
30. **(b)** soft iron
31. **(c)** shorted
32. **(b)** 10 ohm resistor
33. **(c)** conductivity
34. **(b)** thermocouple
35. **(d)** induction exceeds capacitance
36. **(b)** zero
37. **(b)** electron flow
38. **(b)** three-phase
39. **(a)** threads per inch
40. **(c)** shunt
41. **(a)** magnetism
42. **(c)** voltage
43. **(c)** vacuum
44. **(c)** effective
45. **(b)** rheostat
46. **(a)** direct
47. **(c)** 3/4" 346-7b
48. **(d)** commutator
49. **(c)** zero
50. **(c)** good conductors

1. **(d)** NFPA
2. **(d)** plan
3. **(d)** reached quickly DEF 100
4. **(d)** mech function DEF 100
5. **(d)** unbalanced
6. **(d)** I, II or III DEF 100
7. **(d)** hazards DEF 100
8. **(b)** free of shorts 110-7
9. **(a)** Type A ballast
10. **(b)** 1/120
11. **(a)** real power
12. **(c)** demand factor DEF 100
13. **(c)** authority 90-4
14. **(d)** continuously DEF 100
15. **(a)** remote-control DEF 100
16. **(c)** grounded 110-16a
17. **(b)** three-phase 4-wire
18. **(d)** broken
19. **(b)** greater than
20. **(d)** stationary 550-2
21. **(c)** an outlet DEF 100
22. **(a)** thermal protector DEF 100
23. **(a)** instructions 110-3b
24. **(a)** free from hazard 90-1b
25. **(d)** output of motor

26. **(d)** I, II, & III
27. **(a)** one wattmeter
28. **(d)** I, II & III
29. **(c)** chapter 7
30. **(d)** temperature
31. **(c)** continuity
32. **(a)** in series with the load
33. **(c)** at least two wires
34. **(d)** wire binding screws
35. **(c)** fine print notes
36. **(c)** both I & II DEF 100
37. **(d)** I, II & III DEF 100
38. **(d)** continuous <u>DUTY</u> DEF 100
39. **(c)** dust will not interfere DEF 100
40. **(b)** DEF 100
41. **(a)** guarded DEF 100
42. **(c)** service drop DEF 100
43. **(a)** dead front DEF 100
44. **(c)** is a constant value
45. **(d)** <u>ampacity</u> remains the same
46. **(d)** I, II or III
47. **(b)** #18 402-6
48. **(a)** watt output decreases
49. **(b)**
50. **(c)** antenna systems 820-1

1. **(a)** expansion joints
2. **(b)** interrupting
3. **(d)** multiply strands
4. **(d)** effective value
5. **(b)** commutator bar separators
6. **(d)** not to protect the end of wire
7. **(a)** tin & lead
8. **(d)** safe for stepping down
9. **(d)** XC
10. **(b)** electrical energy only
11. **(c)** insulation resistance
12. **(d)** controller DEF 100
13. **(d)** longer bulb life
14. **(c)** apparent
15. **(b)** 480v
16. **(c)** all in parallel
17. **(b)** voltage difference
18. **(b)** current the same in series
19. **(b)** moisture proofing
20. **(b)** prevent corrosion
21. **(c)** ammeter, wattmeter, voltmeter
22. **(c)** low resist. & low melting point
23. **(d)** all DEF 100
24. **(a)** resistor
25. **(c)** asbestos

26. **(a)** fuse clips are too loose
27. **(c)** fitting
28. **(b)** inductance
29. **(c)** effective difference
30. **(c)** turns-ratio
31. **(c)** specific gravity
32. **(a)** magnetic effect
33. **(d)** 1,000,000
34. **(c)** loose connection
35. **(d)** switch
36. **(a)** not permanently enclosed DEF 100
37. **(c)** farad
38. **(c)** even spacing, numerous lights
39. **(c)** one hot leg is shut off
40. **(a)** one
41. **(d)** lightning arresters
42. **(c)** transformer
43. **(b)** oscilloscope
44. **(b)** aluminum
45. **(a)** 90°
46. **(a)** XL
47. **(a)** insulation not the only protection
48. **(d)** polarity
49. **(b)** R/Z
50. **(d)** CO_2 fire extinguisher

1. **(d)** capacitance exceeds inductance
2. **(b)** thread is tapered
3. **(d)** either I or III 230-22
4. **(a)** commutator
5. **(b)** cable terminal
6. **(d)** 70.7v 100v x .707 = 70.7v
7. **(c)** raintight & arranged to drain
8. **(a)** one strand of the conductor
9. **(d)** highest when bulb is on
10. **(a)** remote-control
11. **(c)** 1/2 ohm
12. **(d)** I and II
13. **(b)** shorted
14. **(a)** leather gloves over rubber
15. **(a)** cartridge fuses
16. **(c)** same ampacity T. 310-16
17. **(a)** accidentally energized
18. **(d)** I, II & III
19. **(d)** electrolyte
20. **(c)** DC
21. **(c)** connect two sections
22. **(d)** rawl plugs
23. **(c)** B - C - A
24. **(d)** lack of isolation
25. **(c)** I, II & V

26. **(a)** device DEF 100
27. **(c)** reduce pitting of contacts
28. **(b)** quench the arc
29. **(a)** inversely proportional to csa
30. **(c)** frequency meter
31. **(a)** chemical
32. **(b)** henrys
33. **(a)** DC motor
34. **(a)** some sort of ballast
35. **(b)** less
36. **(a)** 40° C
37. **(c)** 120°
38. **(a)** conductor moving inside a magnetic
39. **(a)** approved sealing compound
40. **(d)** 1/240 second
41. **(c)** watt-hour
42. **(a)** mica
43. **(c)** series-parallel
44. **(c)** eddy current loss at minimum
45. **(a)** downward
46. **(a)** impedance
47. **(a)** squirrel cage
48. **(a)** $R = V^2/W$
49. **(d)** 1000 amp turns
50. **(b)** oil

1. **(c)** increase in vd
2. **(a)** tool to bend pipe
3. **(c)** oil a <u>sleeve</u> bearing
4. **(a)** loose conection
5. **(a)** six
6. **(d)** vacuum or gas
7. **(a)** cabinet DEF 100
8. **(b)** grounded cond DEF 100
9. **(a)** DEF 100
10. **(b)** intermittent DEF 100
11. **(b)** receptacle outlet DEF 100
12. **(c)** covered DEF 100
13. **(c)** covered by Code 110-13b
14. **(b)** authority DEF 100
15. **(a)** DEF 100
16. **(b)** chapter 5
17. **(a)** yes 300-3c1
18. **(a)** load
19. **(c)** RHW
20. **(a)** 10%
21. **(a)** reduce 240-11
22. **(d)** service DEF 100
23. **(b)** DEF 100
24. **(a)** continuous <u>load</u> DEF 100
25. **(d)** the contact resistance

26. **(b)** piezoelectricity
27. **(b)** higher voltage
28. **(c)** rectifier
29. **(a)** delta
30. **(c)** III & IV only
31. **(d)** 1000 W = I^2R
32. **(b)** will not
33. **(d)** high
34. **(c)** compensate for vd
35. **(c)** largest
36. **(b)** main bonding jumper DEF 100
37. **(a)** good connection 110-14a
38. **(d)** megger
39. **(a)** protect user
40. **(a)** control speed of motor
41. **(d)** ammeter, wattmeter & voltmeter
42. **(a)** 25 or 60 cycle
43. **(b)** blades dead
44. **(b)** lead sheath
45. **(d)** damaging the finish
46. **(b)** rubber
47. **(d)** low wattage-high resistance
48. **(b)** commutator or slip rings
49. **(c)** 6"
50. **(d)** prevent electrical shock

1. **(b)** Class I Div. II 511-3a
2. **(a)** 6' 250-83,84
3. **(b)** 75a 430-110a
4. **(d)** not at all 370-10
5. **(d)** none of these 650-7
6. **(d)** I,II,III & IV 450-11
7. **(c)** 100% 210-21b1
8. **(d)** 6' 370-18b
9. **(b)** discharging 280-2
10. **(d)** a & c 620-11a
11. **(c)** 6' 250-91b ex.1a
12. **(d)** 115% 440-12a1
13. **(c)** neutral 250-93a 445-4d
14. **(a)** Class I Div.I 516-2a5
15. **(a)** 90 days 305-3b
16. **(d)** removed 354-7
17. **(a)** line side 700-12e
18. **(d)** all 410-24a
19. **(a)** #14 600-5 ex.2
20. **(a)** ozone 310-6
21. **(d)** any 210-5b
22. **(a)** 300v 352-1,22
23. **(a)** 50% momentary 660-5
24. **(c)** Class III Div. II 500-7b
25. **(b)** simul. 380-2b ex.1 240-22 ex.1
26. **(c)** 12' 210-62
27. **(c)** 50° 424-36
28. **(c)** 800a 240-3 ex.4
29. **(c)** GFP 230-95b FPN 2
30. **(c)** approved 410-31 ex.1
31. **(c)** 90% 555-5
32. **(d)** stranded 410-28e
33. **(b)** 660w 720-5
34. **(c)** 10v 517-64a1
35. **(a)** air break 230-204a

36. **(b)** corrosion prot. 346-1c
37. **(b)** 4" 550-10h
38. **(b)** 8" 342-7a2
39. **(c)** protect 90-1a
40. **(d)** I,II & III 320-6,16 380-10a
41. **(a)** AC 770-6a,b,c
42. **(b)** service entrance 250-79d
43. **(a)** 12" 450-22
44. **(d)** 0 T. 350-3
45. **(d)** I,II & III 339-1a
46. **(d)** liquid-tight 502-4a2
47. **(c)** time delay 440-54b
48. **(d)** not allowed 502-2a3
49. **(a)** isolating 380-13a
50. **(d)** GFCI 680-31
51. **(a)** 3' T. 110-16a
52. **(d)** .8 Chapter 9 Table 1 Note 3
53. **(c)** not exceed 600v 430-111c
54. **(b)** #6 280-23
55. **(c)** 2" 600-8d
56. **(c)** 35a 50a x .707 = 35a
57. **(b)** II only 240-8 ex.
58. **(a)** 1/4" 600-33c
59. **(b)** 115% 445-5
60. **(b)** fixed 400-8
61. **(c)** 3/4" 250-83c1
62. **(d)** 125% 520-25a
63. **(b)** MI 518-4
64. **(d)** all 364-28
65. **(b)** 1/4" 300-6c
66. **(c)** 100a 430-72b ex.1 T.430-72b Col.Bex
67. **(b)** hoistways 318-4
68. **(b)** 24" 514-8 ex.2
69. **(c)** #8 310-3
70. **(d)** 100a 550-23b

1. **(c)** 6' 410-67c
2. **(c)** 2 sq.ft. 250-83d
3. **(d)** I, II & III 430-82b
4. **(c)** 12 1/2% 250-79c
5. **(b)** ampacity 430-62b
6. **(b)** #10 T.250-95
7. **(b)** elect. passing a point
8. **(b)** 30a 240-51a
9. **(d)** both 440-41b
10. **(c)** a & b 210-4 ex.1,2
11. **(d)** I, II & III 600-5
12. **(b)** tension 400-10
13. **(d)** none T. 220-3b
14. **(c)** 125% 645-5a
15. **(c)** locked 430-102b ex.
16. **(a)** grounded 250-7a
17. **(c)** movement 318-8d
18. **(b)** dust 500-6
19. **(b)** header duct 356-1 358-1
20. **(b)** 3-wy & 4-wy 230-84b ex.
21. **(c)** 4' 424-59 FPN
22. **(a)** hazardous 330-3
23. **(c)** 5 1/2' 210-52a
24. **(d)** #17 T.810-16a
25. **(a)** #14 650-5a
26. **(d)** 5/8" 501-5c3
27. **(a)** 3.36" Chapter 9 Table 4 100%
28. **(c)** dry 353-2
29. **(c)** I or III 280-4a
30. **(c)** I, II & IV 680-23b
31. **(d)** all DEF 100
32. **(b)** Class I, Div.II T.515-2
33. **(d)** all 305-4c
34. **(c)** signal 422-12
35. **(c)** III only 250-125

36. **(b)** cleat type 410-3 ex.
37. **(c)** 70% 220-22
38. **(c)** locking 380-6b
39. **(d)** I, III & IV 424-19
40. **(b)** beginning 215-5
41. **(c)** average, it is the RMS effective value
42. **(b)** 12' T. 346-12
43. **(a)** 12" 470-3
44. **(c)** 15' 225-6b DEF
45. **(c)** electric-discharge 600-7a
46. **(c)** watts 430-7
47. **(c)** 0.030a 760-6 ex.
48. **(c)** 36" 328-10
49. **(b)** II only 670-2
50. **(c)** 60°F 230-24
51. **(d)** I, II & III 410-51
52. **(d)** 75% 362-6
53. **(d)** continuous 620-61b2
54. **(a)** MI 330-3
55. **(c)** #10 810-21h
56. **(d)** I,II,III & IV T.430-152
57. **(b)** .0327 Chapter 9 Table 5
58. **(b)** 6' 250-79f
59. **(a)** #26 640-7
60. **(b)** #18 402-6 410-24b
61. **(b)** galvanic action 346-1b
62. **(d)** a & b 427-2 FPN
63. **(c)** GFCI 600-11
64. **(d)** electric furnace 250-5b4 ex.1
65. **(b)** 1" 374-7
66. **(a)** one more 680-21d
67. **(a)** energized 300-31
68. **(a)** mogul 410-53
69. **(c)** electric-discharge 220-22
70. **(a)** 24" T.300-5

1. **(d)** 1/4" 373-2
2. **(d)** not required 370-13d
3. **(a)** within sight 600-2a
4. **(b)** 310-13 FPN
5. **(d)** 354-2
6. **(b)** release-type 328-10
7. **(b)** 300-22c
8. **(c)** racked 110-12b
9. **(a)** 65a Note 8 ex.4
10. **(c)** 7 430-9c
11. **(c)** 2 x 4 410-15b1
12. **(a)** both 351-4a
13. **(d)** four 517-18b
14. **(c)** both ends 280-25
15. **(b)** 16 551-2
16. **(d)** 15 424-99b
17. **(b)** accidental 430-73
18. **(d)** I or II 430-109
19. **(b)** 0.06 250-83d
20. **(d)** either 460-28b
21. **(a)** 249a 220-22
22. **(d)** neither 250-114
23. **(d)** neither 501-9a3
24. **(c)** panelboard 384-16 ex.1
25. **(b)** cable trays 340, 4,5
26. **(d)** one minute 701-11
27. **(c)** both 250-51
28. **(c)** 6 240-100
29. **(c)** 24 T.710-3b ex.1
30. **(b)** II only 517-72c
31. **(b)** II only 250-42 ex.3
32. **(c)** 30 374-2
33. **(b)** current-carrying 520-53n
34. **(a)** potenial diff. 250-86 FPN2
35. **(b)** vented power fuse DEF 100

36. **(a)** 600 331-4 (6)
37. **(d)** 2' 550-23e
38. **(b)** below 210-52a FPN
39. **(d)** I,II,III 511-10
40. **(b)** readily accessible 450-13
41. **(d)** cold water pipe 250-81a
42. **(d)** 4 1/2' 336-15
43. **(a)** b.c. rating 220-3a
44. **(b)** simultaneously 410-48
45. **(d)** 1000 517-15
46. **(b)** frames insulated 250-42f ex.2
47. **(a)** I only 354-3a
48. **(d)** yellow stripe 200-6d
49. **(c)** both I & II 501-8a
50. **(d)** 225 600-10a
51. **(c)** both I & II 250-81a
52. **(c)** 7" T. 710-33
53. **(d)** I, II or III 250-45d FPN
54. **(a)** 18" 760-28b1
55. **(d)** #12 720-4
56. **(b)** 194° F 333-5 ex.
57. **(d)** barrier 810-18c
58. **(b)** interchangeable 410-105
59. **(b)** same grd. electrode 250-54
60. **(b)** 300v 380-8b
61. **(d)** 3 T.430-37
62. **(a)** #14 310-14
63. **(c)** 12' 511-7b
64. **(d)** arranged in cable 230-30 ex.d
65. **(c)** both I & II 430-125 c1a
66. **(d)** 200% 630-12a
67. **(d)** neither I nor II 501-16a
68. **(c)** both I & II 250-23a
69. **(a)** 31% Chapter 9 Table 1
70. **(d)** I,II or III 250-71b1,2,3

1. **(a)**   370-7c ex.
2. **(d)**   all of the above 230-30 ex.
3. **(c)**   hot water piping 800-40b1
4. **(a)**   6 1/2' 210-8a3 FPN
5. **(d)**   I,II or III 250-79a
6. **(a)**   equip. grd. 250-51
7. **(d)**   30 530-14
8. **(c)**   bottom shield 328-2
9. **(a)**   comb. dust 502-1
10. **(b)**   200   110-16f ex.
11. **(c)**   both 675-15 250-46
12. **(c)**   I or II 422-23
13. **(c)**   dist. marked 250-42 ex.4 250-45 ex.3
14. **(d)**   36.96 T. 610-14a
15. **(a)**   grd. cond. 200-3
16. **(d)**   I, II & III 700-12b
17. **(d)**   indoor port. 600-32b
18. **(a)**   recpt. outlet 210-50a
19. **(c)**   III only 410-54a
20. **(b)**   400% 530-18a
21. **(a)**   100v 640-5
22. **(b)**   splices & taps 347-16
23. **(b)**   terminals 110-14a
24. **(d)**   I, II & III 336-16
25. **(a)**   4 517-2
26. **(b)**   fountains 680-51a
27. **(b)**   300 630-32a
28. **(b)**   5 330-13
29. **(a)**   #6 230-202a
30. **(d)**   csa 45% 356-5
31. **(d)**   I,II, or III 710-3b
32. **(c)**   125 690-8a
33. **(b)**   20a or less 250-91b ex.1b
34. **(c)**   384-15
35. **(b)**   II only 550-5a

36. **(c)**   351-9 ex.2
37. **(d)**   16 T.346-12
38. **(d)**   none 250-91a
39. **(a)**   3/4" T.384-36
40. **(b)**   II only 440-2
41. **(d)**   24" Chapter 9 note 3
42. **(c)**   #6 250-94 ex.1a
43. **(b)**   readily accessible 240-10
44. **(c)**   2.15 318-11 b4
45. **(a)**   I & II only 210-8a 5 & 6
46. **(b)**   48 ... 60 680-41h
47. **(c)**   I or II 460-25c
48. **(a)**   35 422-18
49. **(c)**   12 1/2% 250-23b
50. **(d)**   1 1/2" 410-76b
51. **(c)**   75 210-63
52. **(c)**   divided equally 410-102
53. **(c)**   waves 555-3 FPN3
54. **(b)**   511 - 517 510-1
55. **(d)**   source of supply 230-204d
56. **(c)**   I or III only 280-4a
57. **(c)**   50 364-11
58. **(d)**   header 358-9
59. **(c)**   both I & II 240-82
60. **(d)**   water cooled 503-6
61. **(d)**   150 424-72a
62. **(c)**   both I & II 430-35a, b ex.
63. **(c)**   7" T. 710-33
64. **(c)**   both I & II 250-23b
65. **(c)**   300 680-51b
66. **(c)**   both I & II 513-2a and c
67. **(b)**   louvered 450-43a
68. **(a)**   trolley frame 610-61
69. **(c)**   both I & II 501-8
70. **(b)**   30 T. 710-3b

1. **(d)** I,II & III 410-16c
2. **(a)** lead covered 310-8a
3. **(c)** disconnect all 210-4b
4. **(c)** individual 450-2
5. **(c)** afford protection 110-12a
6. **(d)** maintained Note 8b
7. **(c)** liquidtight 553-7b
8. **(a)** 600v 225-10
9. **(c)** both I & II 300-20b
10. **(c)** 50% Table 220-13
11. **(c)** authority 555-8
12. **(c)** 1/3 450-5 FPN
13. **(b)** identified 780-3a
14. **(c)** underground 310-15d
15. **(d)** indicate 110-22
16. **(c)** major diameter Chap. 9 Table 1 note 5
17. **(b)** II only 700-26
18. **(a)** twice 220-19
19. **(d)** MV 326-1
20. **(d)** all 430-32d
21. **(c)** 250% 450-6b
22. **(c)** #8 540-13
23. **(d)** all 600-9a
24. **(a)** 20a 410-56b
25. **(a)** outlet 600-6b
26. **(b)** 30 ma 600-32b
27. **(b)** II only 410-76a
28. **(b)** instrinsically safe 504-4
29. **(b)** suitable 110-8
30. **(d)** 324-5b
31. **(d)** I,II & III 90-2
32. **(d)** lighting & power 110-19
33. **(b)** one in ungrd. T.430-47
34. **(c)** 120 mechanical degrees
35. **(b)** additional prot. 250-5b FPN

36. **(b)** I & II only 700-15
37. **(b)** .1194 Chapter 9 Table 5a
38. **(b)** 20a 422-27e
39. **(d)** 87 1/2% 701-11a
40. **(b)** 1 second 230-95a
41. **(d)** loop connectors 300-19b1,2,3
42. **(b)** car lights 620-22
43. **(b)** 12" Table 300-5
44. **(a)** installed 230-95c
45. **(b)** insulating mats 430-133
46. **(d)** high-voltage 370-52e
47. **(d)** limits access 517-18c
48. **(c)** 6 times 230-208
49. **(d)** hazardous 352-1
50. **(b)** either side 501-5b2
51. **(b)** 3/4" 354-3a
52. **(d)** 6 1/2" T. 373-6b
53. **(c)** #14 820-40a3
54. **(c)** 8' vert. & 5' hor. 250-42a
55. **(b)** raintight 300-6a
56. **(d)** I,II & III 410-57a
57. **(b)** weatherproof 373-2a
58. **(b)** I, II & IV 331-3 1,4,5
59. **(a)** 100% 450-6a2
60. **(c)** 125% 422-4a ex.2
61. **(c)** simultaneously 230-74
62. **(b)** 2 1/2" T. 373-6a
63. **(d)** 150% 445-4c
64. **(d)** I, II & III 330-3
65. **(d)** uninsulated 338-1c
66. **(d)** 6 times 430-110 c3
67. **(c)** shielded 240-41a
68. **(d)** 24" 310-11b1
69. **(d)** 10' 600-9c
70. **(d)** I or II only 440-6a ex.

1. **(b)** same size T. 250-95
2. **(b)** 10° 310-13 FPN 402-3 FPN
3. **(a)** raintight 230-54a
4. **(d)** 24" Table 300-5
5. **(b)** nonhazardous 516-2d
6. **(c)** right angle 358-5
7. **(d)** 2" 501-5a2
8. **(a)** listed 600-4
9. **(c)** 400% 530-18b
10. **(b)** out of reach 230-9ex.
11. **(c)** 20 cells 480-5b
12. **(b)** within sight 600-2a ex.
13. **(c)** temperature rise 470-18e
14. **(c)** 462 amp 555-5
15. **(a)** 5 T. 370-6a
16. **(a)** lighting circuit 700-12f
17. **(c)** both 210-52b ex.1,2
18. **(b)** 300v T.402-3
19. **(d)** #10 225-6a1
20. **(a)** 5' 680-6 a1 ex.
21. **(d)** 20' 250-81c
22. **(c)** flex.metal 501-4b
23. **(a)** 4 1/2' 350-4
24. **(c)** 135% 460-8a
25. **(b)** 645-3
26. **(d)** adjacent T.310-16 Note 5
27. **(b)** 90° 410-5 410-65a
28. **(d)** elect. cont. 250-92b
29. **(d)** nonhazardous 513-2d
30. **(d)** ranges 550-2 FPN
31. **(a)** trans. assembly 328-15
32. **(a)** true 700-7a
33. **(d)** 334-10e
34. **(b)** 2 1/2' 250-81d
35. **(d)** six 517-19b

36. **(b)** 12' 230-24b
37. **(b)** identified 346-15c
38. **(c)** II & III 250-91a
39. **(c)** 15-20-30 424-3a
40. **(b)** continuity 318-6a
41. **(a)** true 545-6 ex.
42. **(c)** therm. protector DEF 100
43. **(c)** 8' 410-4d
44. **(c)** hazard 240-3 ex.3
45. **(c)** #10 110-14a
46. **(a)** substantially increased 300-21
47. **(b)** 20a 600-6a
48. **(b)** 6" 300-14
49. **(d)** soldered 324-12
50. **(b)** 20a 600-6b
51. **(c)** 6" T. 230-51c
52. **(b)** #4 370-18a
53. **(b)** 12" 225-14d
54. **(c)** 1" x 2" 370-13b2
55. **(d)** 3 1/2" 410-38c
56. **(b)** replacement 240-51b
57. **(b)** metal 430-12a
58. **(b)** 1 1/2 hours 700-12a
59. **(c)** 15/16" 370-14
60. **(c)** SNM 337-1
61. **(d)** 10' 440-64
62. **(a)** 5' 374-3
63. **(d)** 4 single or 2 duplex 517-18b
64. **(b)** #4 373-6c
65. **(c)** 5.0 amps T. 725-31a
66. **(b)** 16" 410-15a
67. **(a)** 1 1/4" 300-4a1
68. **(c)** 50v 110-17a
69. **(a)** 65v 445-4c
70. **(a)** 1/2" 370-14

1. **(d)** I,II or III   250-5a1,2,3
2. **(c)** both I & II   250-6b1,2
3. **(c)** 120/240 1 ø   550-21
4. **(d)** I,II & III   280-4b FPN2
5. **(b)** listed   240-20c
6. **(d)** cable tray   318-2
7. **(a)** offset   373-6b FPN
8. **(b)** raceway fitt.   410-16a   370-13d
9. **(c)** both I & II   370-15   370-18b
10. **(d)** neither I nor II   551-20e
11. **(d)** same raceway   310-4
12. **(a)** 2.5"   330-13
13. **(a)** MI   230-202b
14. **(d)** I,II & III   760-11,12,15
15. **(a)** #8   547-8b
16. **(b)** single   427-47
17. **(d)** as far away as possible   450-45a
18. **(b)** insulated   324-4
19. **(c)** GFCI   422-8d3
20. **(b)** 8 times   300-34
21. **(a)** 2
22. **(a)** 8'   250-42a
23. **(b)** 80%   501-8a
24. **(a)** #18   410-27b ex.
25. **(a)** readily identified   700-9a
26. **(c)** readily acc. external   240-6
27. **(b)** appliance   422-15c
28. **(b)** electrostatic   516-5d
29. **(c)** I,III & IV   300-22b
30. **(d)** 120v   170v x .707 = 120v
31. **(c)** written record   700-4d
32. **(c)** both & II   230-81
33. **(d)** I, II & III   410-29 a,b,c
34. **(a)** distinctive   200-6a ex.4
35. **(d)** I, II, & III   700-12b1,2,3

36. **(b)** braces or guys   230-28
37. **(a)** 10'   225-18
38. **(d)** I, II & III   680-24
39. **(b)** 5"   T.430-10b
40. **(a)** internal   90-6
41. **(c)** continuous duty   T.430-22a ex.
42. **(a)** ungrounded   240-20a
43. **(c)** 670-4b
44. **(d)** 100'   620-41
45. **(d)** own section   702
46. **(b)** 90°   110-16a
47. **(d)** I, II & III   640-1
48. **(b)** II only   240-100
49. **(d)** not required   250-3a ex.2
50. **(c)** permanent   517-2 DEF
51. **(a)** phase
52. **(d)** I, II or III   450-25
53. **(c)** I & III only   600-8b
54. **(c)** 300v   410-73b
55. **(c)** I & III only   701-2 FPN
56. **(d)** need <u>not</u>   250-122 ex.
57. **(d)** rigid   346-1c
58. **(b)** metal   358-6
59. **(b)** II only   424-29
60. **(a)** I, II & III   600-22
61. **(d)** same or greater   440-4c FPN
62. **(c)** 0.816   Table 4 40% of 2.04
63. **(d)** 17 1/2"   T.349-20a
64. **(b)** supported   370-13d
65. **(a)** I, II or III   600-36a
66. **(b)** I, II & IV   430-142
67. **(c)** III only   600-21e
68. **(c)** 3.5 kva   Chapter 9 Example 2b
69. **(c)** NFPA   90-5
70. **(d)** 1.34   Chapter 9 Table 4

1. **(b)** single   555-3
2. **(d)** I, II & III   680-22a2,6   680-23a
3. **(d)** shall not   605-5c
4. **(b)** 620-101b
5. **(c)** 3/8"   T. 430-12c1
6. **(a)** .0625"   370-20b
7. **(a)** 3 gallons   460-2a
8. **(c)** accessible only   700-25
9. **(c)** 250-43b ex.   650-4
10. **(c)** III only   318-5e
11. **(d)** 35%   T. 310-16 Note 8
12. **(c)** shielded   400-31b
13. **(c)** 680-25d
14. **(c)** #6   800-40d
15. **(b)** wooden floors   430-14b
16. **(d)** I & II only   600-2
17. **(c)** is covered   640-1
18. **(c)** 6'   T. 347-8
19. **(d)** 200%   630-12b
20. **(c)** 480-2
21. **(d)** I, II & III   702-2 FPN
22. **(a)** #14   675-13
23. **(a)** 250-95 ex.2
24. **(d)** I, II & III   551-2 FPN
25. **(d)** 2000a   374-6
26. **(b)** equip. grd.   515-5c
27. **(b)** terminals   200-10a
28. **(c)** 1"   373-11a3
29. **(c)** 1/2 ohm parallel
30. **(a)** electrolytic cell   668-2
31. **(b)** 8'   300-5d
32. **(d)** #20   620-12a2
33. **(c)** I & III only   610-21f4
34. **(d)** 18"   370-13e
35. **(a)** dressing rooms   520-73

36. **(c)** obelisk   T. 310-16 footnotes
37. **(b)**
38. **(b)** II only   725-38a2
39. **(b)** 48...60   424-72b
40. **(a)** 1"   331-5b
41. **(b)** 8"   547-4
42. **(b)** no heating   300-20b FPN
43. **(b)** 2   210-70 FPN
44. **(b)** 16 1/2w   424-44a
45. **(a)** main bonding jumper   384-3c
46. **(c)** two hours fuel   700-12b2
47. **(b)** 4'   422-8d2a
48. **(c)** III only   215-8   230-56   384-3e,f
49. **(b)** arms & stems   410-28c
50. **(c)** gasoline pump   514-5
51. **(d)** 6'   820-10e3
52. **(d)** I, II & III   250-152b1,2,3
53. **(a)** adequate bonding   250-44 FPN
54. **(c)** I & III only   400-22a
55. **(c)** 200%   600-2b
56. **(c)** shielded   310-6
57. **(b)** 50%   660-6a
58. **(a)** Class I   250-26b ex.
59. **(c)** own section   540
60. **(d)** I, II & III   669-9
61. **(a)** 8'   250-84 FPN
62. **(a)** 1000v   410-80b
63. **(a)** ungrounded   685-12
64. **(a)** bare copper   338-1b
65. **(a)** I only   410-91
66. **(a)** surge arrester   280-2
67. **(c)** overload   430-55
68. **(b)** 168a   T. 325-14
69. **(b)** 125...250   440-55b
70. **(c)** both   354-6

1. **(b)** #20   620-12a1 ex.
2. **(a)** insulated 300-3c
3. **(d)** 35kv 450-21c
4. **(b)** 18" 250-95a
5. **(c)** .040 410-38a
6. **(a)** not req. 600-8a ex.
7. **(a)** 8' T.400-4 note 3
8. **(b)** outside 220-3b
9. **(d)** I, II & III 110-18
10. **(b)** Class II, Div.I 500-6a
11. **(b)** 25a 210-3
12. **(d)** #14 620-12a1
13. **(c)** 600a T.318-7b2
14. **(c)** III only 250-1 FPN
15. **(c)** polarity 200-11
16. **(b)** separately 640-6b
17. **(a)** 1va T.220-3b
18. **(a)** at all 422-5
19. **(c)** I & II 110-11
20. **(a)** true 520-24
21. **(a)** 3" 384-10
22. **(c)** I, II & III 230-50a
23. **(b)** 100 518-1
24. **(b)** #4 250-92a
25. **(d)** both 700-12a
26. **(a)** D 310-11c
27. **(b)** #14 410-51
28. **(c)** pigtail 300-13b
29. **(d)** grd. electrode DEF 100
30. **(c)** 300% 430-72b ex.2
31. **(d)** V 336-25b
32. **(d)** none 410-8d3
33. **(c)** 500mv 517-15a
34. **(b)** unbalanced 220-22
35. **(d)** 48 times 370-51a
36. **(c)** bonding jumper 250-77
37. **(b)** ex. moisture 501-5
38. **(a)** 8 384-15
39. **(c)** 3" 810-54
40. **(a)** 6" Table 300-5
41. **(b)** simult. 514-5
42. **(c)** close to wall 210-52
43. **(a)** rigid metal 501-4a
44. **(b)** 3 circuits Chapter 9 example 1a
45. **(d)** Class I, Div.I 511-3b
46. **(d)** I, II & III 250-117b
47. **(d)** 3 hours 450-43
48. **(c)** source of heat 410-54a
49. **(c)** 20a 210-19b ex.2
50. **(b)** 8.19a 630-31a2
51. **(d)** all 310-11a1,2,3
52. **(d)** 75% 640-4 ex.a
53. **(c)** similar change 240-23
54. **(a)** gases & vapors 500-5
55. **(b)** 15a 363-10
56. **(d)** any 600-22 410-23,27
57. **(d)** #8 230-31b
58. **(a)** 18" 501-5a1
59. **(c)** 3/4" 346-7b
60. **(b)** grounding cond. DEF 100
61. **(d)** 6 1/2' 380-8
62. **(a)** true 110-9
63. **(b)** will not 230-95 FPN4
64. **(a)** general-use type 501-6b2
65. **(c)** both 336-10b 333-12
66. **(d)** Class I, Group C 517-61a5
67. **(b)** 2' 333-11 ex.1
68. **(a)** Class I, Div.I 517-60a2
69. **(b)** 30a 600-6a
70. **(a)** identified 230-28

1. **(c)** 1/2" 410-46
2. **(b)** 208v 424-35
3. **(b)** 6' 210-50c
4. **(d)** cabinet DEF 100
5. **(d)** 25' 250-33 ex.1
6. **(c)** both 230-54f
7. **(b)** 0.027" Chapter 9 Table 8
8. **(c)** 75a 240-6
9. **(b)** enclosed 600-34e
10. **(c)** II, III and IV 630
11. **(c)** carpet squares 328-1
12. **(c)** 2 1/2' 230-51a
13. **(b)** started 430-7b1
14. **(a)** 35' 810-11 ex.
15. **(a)** red 424-35
16. **(a)** 18' 680-4
17. **(c)** I, II & IV T.310-13 410-31
18. **(b)** exposed to weather 600-21c
19. **(c)** #18 or larger 240-4 ex.1
20. **(c)** 50a 422-28c
21. **(c)** end seal T.310-16 Note 6
22. **(b)** clothes closets 240-24d
23. **(b)** 48" 230-54 ex. = 2 x 24" = 48"
24. **(a)** 1/2" to 4" 351-5a,b
25. **(d)** I,II,III & IV 630-31b FPN
26. **(d)** 200a 380-16
27. **(c)** 18' 210-6d 1b
28. **(a)** tamperproof 517-18c
29. **(b)** 50a T.310-16 Note 8 ex.3
30. **(b)** MTW T. 310-13
31. **(d)** EMT 502-4
32. **(b)** 10 seconds 700-12
33. **(d)** I, II & III 250-92a
34. **(a)** RMS 620-13 FPN
35. **(c)** flame arrestor 480-9a

36. **(a)** are not 517-30b
37. **(d)** not classified 511-2
38. **(b)** 6 pounds 410-15a
39. **(c)** solar cell 690-2
40. **(a)** 90-2a1
41. **(b)** 5000v 600-32b
42. **(d)** on ceiling 410-8a1,2
43. **(a)** 10' 225-4
44. **(d)** all 400-10 FPN
45. **(d)** either 230-22
46. **(c)** #10 517-14
47. **(b)** 31" T. 346-10
48. **(c)** flex. metal 502-4a,b 350-2
49. **(d)** I, II & III 328-11
50. **(d)** 150v 520-25c
51. **(c)** 600v 710-54b
52. **(b)** within 511-6a
53. **(d)** none 352-5
54. **(b)** .75 630-11a
55. **(c)** freedom from hazard 90-1b
56. **(b)** 5' 364-5
57. **(c)** III only 680-22b
58. **(d)** inductive current 300-20a
59. **(a)** 50w 422-8a
60. **(c)** free circulation 110-13b
61. **(c)** III only 430-71
62. **(d)** 100' 240-21 ex.10
63. **(d)** 550-11a1
64. **(d)** 30.8a x 208v x 1.732 = 11,096
65. **(c)** 10" 384-10
66. **(d)** #12 680-25b1
67. **(a)** 600v 250-152a1
68. **(b)** 167°F T.402-3
69. **(d)** none of these 250-81
70. **(a)** 15a 384-32

...Prepare for your Electrical Exam with these popular study-aid books!!

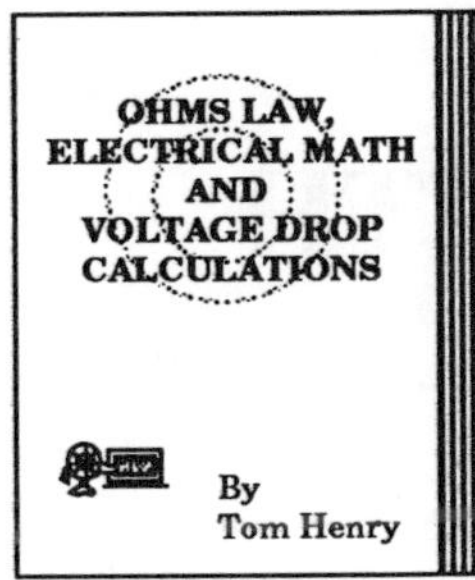

Book #101 - An ideal book for any electrician needing a refresher on Ohm's Law. Explains AC and DC in layman terms, with easier to understand formulas, sketches of circuits, series & parallel circuits, Exact K for voltage drop calculations, function of neutral, review on math for the electrician! Contains over 100 Questions & Answers.

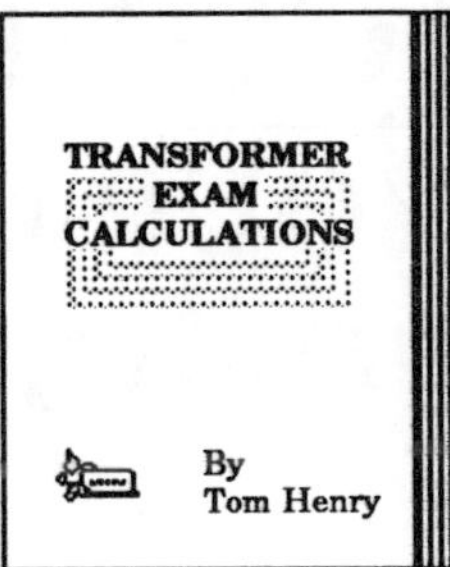

Book #104 - Finally a book written by an electrician in an easy to study format to prepare the everyday electrician in this difficult area of the exam. Single-phase, three-phase, delta-wye, load balancing, neutral calculations, open-delta, high-leg delta, etc. Over 100 calculations with answers!!

Book #102 - An excellent study-aid for the helper, apprentice, or electrician to prepare for the Journeyman license exam. The book contains **8** closed book exams and **12** open book exams. Over **1000** actual exam questions with answers and Code references. An excellent book to study the Code!!

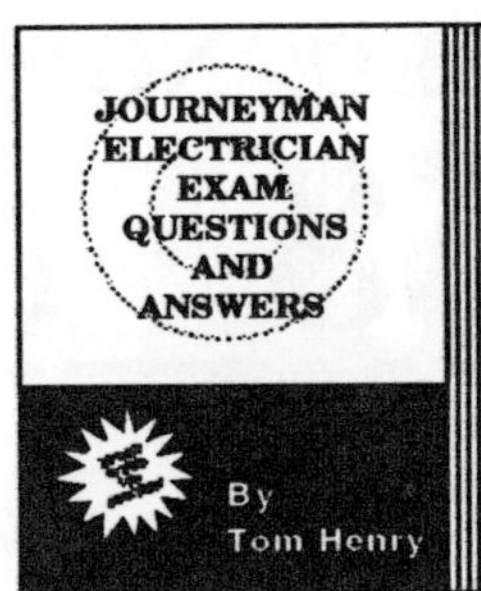

Book #105 - Tom Henry's TOP SELLER!! Everything on calculations- **8** chapters - Cooking equipment demands single-phase ranges on a 3 phase system, ampacity, box-conduit fill, motor circuits, service sizing, feeder sizing, cable tray calculations, and mobile homes, etc. A must!

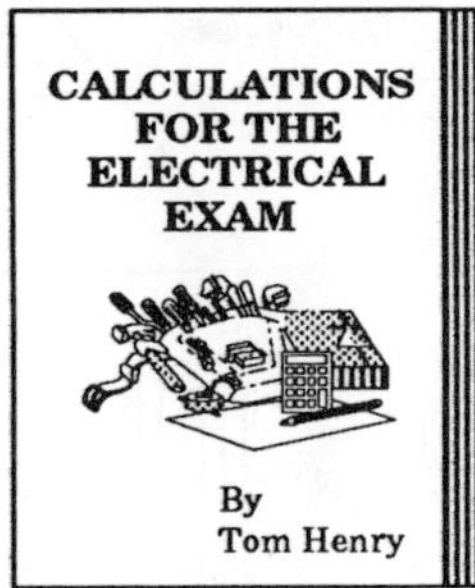

Book #103 - A book designed to advance the electrician in the Code book from the Journeyman level. Contains **6** closed book exams and **10** open book exams. Over **1000** actual exam questions with answers and Code references. An excellent study-aid, takes you cover to cover in the Code including exceptions and Fine Print Notes!!

Book #115 - Tom Henry's book to prepare you for the burglar and fire alarm exam. Hundreds of exam questions with answers. OSHA, UL, NFPA, Business Law, loop circuits, etc. !!!!!!

Book # 110 - Tom Henry's favorite reference book. A complete reference book for the electrician that gives the definitions of the language used in the construction field. Also contains formulas used for the exam and in the field, diagrams showing motor, transformer, and switch connections, etc...

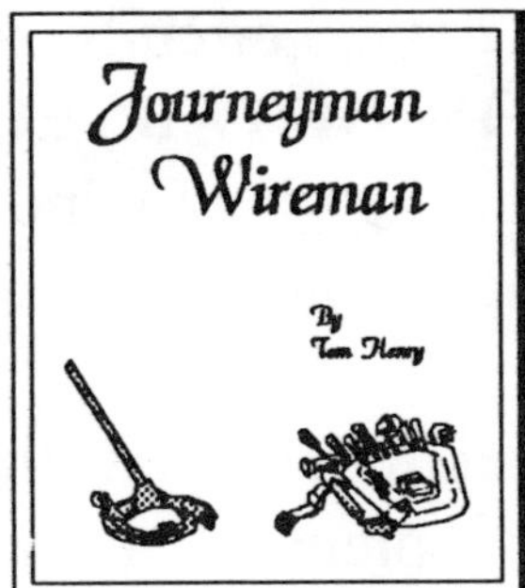

Book # 171 - Tom Henry's quiz book! 60 quizzes on tool identification, wiring methods, blueprint symbols, meter reading, circuit testing, controls, proper installation, etc. This book was written to help prepare an electrician for the mechanical, comprehension and aptitude testing of the exam.

Code Electrical Classes & Bookstore 6832 Hanging Moss Road Orlando, FL 32807
1-800-642-2633

CODE - TABS
INDEX YOUR CODE BOOK WITH
Tom Henry's designed format.

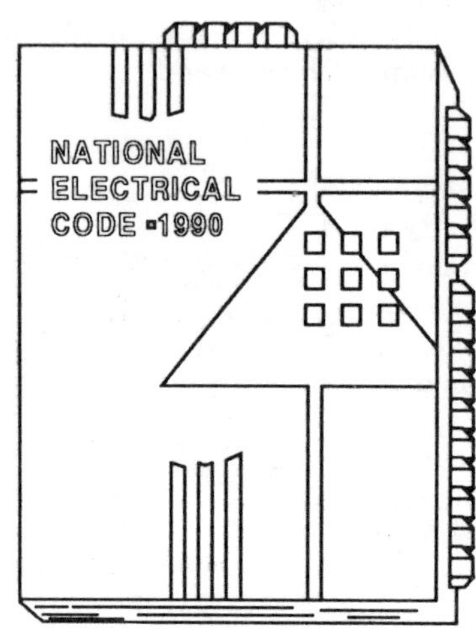

REVISED - NOW CONTAINS

68 TABS

- Tabs to index your Code book! A designed format by Tom Henry to save time and costly errors.
- Have all the **KEY** Code references at your finger tips!
- A special row of service calculation tabs for both residential and commercial. This special row of tabs remind you of all the demand factors that can be applied to sizing the service conductor load.
- 6 motor calculation tabs to size the wire, heaters, breakers, feeders, etc. to motors
- **KEY** tabs for standard size fuses and breakers, sizing equipment grounding conductors, grounding electrode conductors, multi-family optional calculation, burial depths, pools, Class 1, 2 and 3 circuits, signs, welders, hazardous locations, etc.
- **A BRAND NEW ROW OF TABS FOR INDEXING QUICKLY!** The toughest part of an electrical exam is trying to find the answer in the Code book in the allowed time. Now **14 brand new tabs** to aid you in using the **"INDEX"** faster saving the valuable allowed time.
- These tabs are designed for everyone that use the Code book, inspectors, designers, engineers, electricians, apprentices, instructors, helpers, etc.......................................

Tom Henry's **BRAND NEW BOOK!!**

"KEY WORD INDEX"

"Where does it say that in the code book" ??

Now you'll know "where it says that in the code book"!

The KEY WORD INDEX shows the CODE REFERENCE and the **PAGE NUMBER** for the Softback and looseleaf Code books!

Tom Henry has put together the big **TIME SAVER** index to use with your electrical code book.   Find words in the book in **seconds!**

Troublesome words and phrases used in the electrical field are now at your fingertips!   Over 3500 contained in this excellent index!

Have you found it difficult to locate words in the code book, such as: meltingpoint, auxiliary rectifiers, potential transformers, potheads, paddle fans, receptacle outlet, canopy switches, dwelling open-circuit voltage, circuit breaker height, braces or guys, hysteresis, cleat-type lampholders, bus bar ampacity, multi-conductor portable cables, tube keyers, bleeder resistors, pilot lights, output leads, car house, trolley wires, taper per foot, crowfeet, cutouts & flashers, derangement, end seal, hickeys, galvanic action, howlers, linear feet, linoleum, offset, one-shot bender, opposite polarity, ozone-resistant, plaster rings, potting compound, scuttle hole, smoothing iron, tunnels, yokes, wharfs, zig-zag transformers, etc..

ITEM #162 KEY WORD INDEX $10.00

PLACE YOUR ORDER NOW!!

WHEN IN ORLANDO, STOP BY AND VISIT OUR BOOK STORE!

Hours:
8 am to 6 pm

We Carry A Complete Line of Electrical Books

We Teach Electrical Exam Preparation Classes in 18 States.

Call (407) 671-0020 For Information